THE MATTER OF REVOLUTION

THE
MATTER
OF REVOLUTION

Science, Poetry, and Politics
in the Age of Milton

JOHN ROGERS

Cornell University Press   *Ithaca and London*

Parts of Chapter 2 appeared in "The Great Work of Time: Marvell's Pastoral Historiography," reprinted from *On the Celebrated and Neglected Poems of Andrew Marvell*, edited by Claude J. Summers and Ted-Larry Pebworth, by permission of the University of Missouri Press. Copyright © 1992 by the Curators of the University of Missouri. An earlier version of Chapter 5 appeared as "Milton and the Mysterious Terms of History," *ELH* 57 (1990). Reprinted by permission of the Johns Hopkins University Press.

First published 1996 by Cornell University Press.
First printing, Cornell Paperbacks, 1998.

Printed in the United States of America

Library of Congress Cataloging-in-Publication Data
Rogers, John, b. 1961
    The matter of revolution : science, poetry, and politics in the
Age of Milton / John Rogers.
      p. cm.
    Includes bibliographical references and index.
    ISBN 0-8014-3238-3 (cloth: alk. paper)
    ISBN 0-8014-8525-8 (pbk: alk. paper)
    1. English literature—Early modern, 1500–1700—History and
criticism. 2. Great Britain—History—Puritan Revolution,
1642–1660—Historiography. 3. Politics and literature—Great
Britain—History—17th century. 4. Literature and science—Great
Britain—History—17th century. 5. Newcastle, Margaret Cavendish,
Duchess of, 1624?–1674—Criticism and interpretation. 6. Harvey, William,
1578–1657. 7. Marvell, Andrew, 1621–1678—Criticism and interpretation.
8. Milton, John, 1608–1674—Contemporary England. 9. Milton, John, 1608–1674.
Paradise lost. 10. Winstanley, Gerrard, b. 1609. I. Title.
PR435.R64 1996   96-3856
820.9'358—dc20

Cloth printing   10 9 8 7 6 5 4 3 2
Paperback printing   10 9 8 7 6 5 4 3 2 1

In memory of Kipp Rogers
1957–1990

CONTENTS

PREFACE

This book is a study of the cultural intersections between those two events of seventeenth-century history known to us as the English and the Scientific Revolutions. It concentrates specifically on the work of five writers from the period of the Civil Wars, the Interregnum, and the earliest years of the Stuart Restoration: John Milton, Andrew Marvell, Gerrard Winstanley, William Harvey, and Margaret Cavendish. In examining the texts they composed between 1649 and 1666, I investigate the literary and ideological implications of a cultural phenomenon notable for, among other things, its logical out-landishness: the intellectual imperative to forge an ontological connection between physical motion and political action. Although their work spans the generic spectrum from medical treatise to epic poem, each of these writers, I argue, struggles to reconcile the new materialist science of corpuscular motion and interaction and the new political philosophy of popular sover-eignty and consensus. The matter of the English Revolution was for all these figures a problem to be explored, in some cases exclusively, by means of the revolutionary science of matter.

It is the exciting and productive intellectual consequences of this discur-sive commingling at a critical point in England's political and literary history with which this book is concerned. The methodological assumptions behind my analysis of the alliance between political and natural philosophy should be distinguished here at the outset from those informing the study of sev-enteenth-century natural philosophy generated and influenced by the recent work of Steven Shapin and Simon Schaffer.[1] Embarking on what they call a

1. Steven Shapin and Simon Schaffer, *Leviathan and the Air-Pump: Hobbes, Boyle, and the*

"sociology of knowledge," Shapin and Schaffer attempt to isolate the political import of seventeenth-century science in the controversies over experimental method, institutional affiliation, and evidentiary procedure. But a full understanding of the political valences of seventeenth-century science will never arise from a sociology of scientific practice or a positive identification of the allegiances of philosophers and institutions. In limiting their focus to the cultural authority they find subtending certain intellectual institutions and rhetorical methods, the sociologists of early modern science are compelled to overlook the ideological burden borne by the actual physical doctrines over whose success or failure the period's intellectuals waged such battle. For many natural philosophers writing at midcentury, the political promise, or threat, posed by nearly any physical theory—whether vitalist, dualist, materialist, or one of the many combinations thereof—rarely involved the peripheral, if interesting, question of right method. The widely felt social and political implications of the period's scientific speculation emerged much more directly, I am convinced, from an engagement with the ideologically resonant language constitutive of physical theory itself.[2] Reflective of course of my own training as a reader of literary texts, this book makes sense of revolutionary culture through a literary analysis of ideologically consequential trope, narrative, and argument.

The thrust of this literary analysis diverges as well from what I see as the dominant strains of the literary study of the seventeenth-century texts I explore. The distance here from other interpretations of seventeenth-century literary culture can be measured most clearly by my analysis of the poet and political theorist whose name my subtitle employs to stand, synecdochically, for his entire age. The readers of Milton who have pursued the natural philosophy and theology of *Paradise Lost* and *De doctrina christiana* have by and large been much more comfortable describing than actually account-

---

Experimental Life (Princeton: Princeton University Press, 1985). For a related study of the social calibrations of epistemology and method, see Richard W. F. Kroll, *The Material Word: Literate Culture in the Restoration and Early Eighteenth Century* (Baltimore: Johns Hopkins University Press, 1991).

2. I have found more congenial to my own approach the histories of science that have studied the politics of natural philosophy with an eye to the analogical rhetoric of physical explanation, including Carolyn Merchant, *The Death of Nature: Women, Ecology, and the Scientific Revolution* (San Francisco: Harper and Row, 1980); James R. Jacob and Margaret C. Jacob, "The Anglican Origins of Modern Science: The Metaphysical Foundations of the Whig Constitution," *Isis* 71 (1980): 251–67; Otto Mayr, *Authority, Liberty, and Automatic Machinery in Early Modern Europe* (Baltimore: Johns Hopkins University Press, 1986); and two early works by Steven Shapin, "Social Uses of Science," in *The Ferment of Knowledge: Studies in the Historiography of Eighteenth-Century Science*, ed. G. S. Rousseau and Roy Porter (Cambridge: Cambridge University Press, 1980), pp. 95–139, and "Of Gods and Kings: Natural Philosophy and Politics in the Leibniz-Clarke Disputes," *Isis* 72 (1981): 187–215.

ing for Milton's shocking embrace of a radical Christian materialism.[3] Milton's theory of the *ex deo* Creation, his monistic belief in the inseparability of body and spirit, his mortalist belief that the soul dies with the body, and his subordinationist faith that the Son was generated materially from the body of the Father—all these great theological heresies situate Milton so far outside the reaches of either the Puritan or Anglican mainstream that critics have had difficulty placing his theology in a meaningful cultural context. The exceptions to this critical failing, however, are notable, and I must announce here my debt to the work of Christopher Kendrick, Stephen Fallon, and especially William Kerrigan, readers who have, in very different ways, plotted Milton's materialism within compelling critical narratives of motive and function.[4] I have sought in this book to further their expansion of the discursive territory from which the period's most important texts—whether lyric poem, biblical epic, or physical treatise—can be seen to have emerged. My work diverges from Fallon's and Kerrigan's in my attention to the diffused but powerful current of political energy charging the literature of materialism. The weakness of the historical, or "sociological," study of seventeenth-century science has without question been its deafness to the analogical poetics of ontological speculation. The weakness, conversely, of even the strongest literary studies of Milton and his contemporaries has been an insensitivity to the ideological freight of their materialist poetics and theology.

This book opens, in Chapter 1, with an attempt to identify a brief but notable intellectual movement that flourished in the early years of the republican Commonwealth. Citing the example of William Harvey's late writings on the circulation of the blood, I argue that the philosophy of monistic vitalism, which would have such an impact on Milton, emerged in this period to provide a conceptual framework for that social and political structure of self-determination we recognize as liberalism. Turning from the Harveian treatise to the Marvellian pastoral, the next two chapters study the lyric

3. Classic studies of the materialist aspect of Milton's theology include Walter Clyde Curry, *Milton's Ontology, Cosmogony and Physics* (Lexington: University Press of Kentucky, 1957); William B. Hunter, Jr., "Milton's Materialistic Life Principle," *Journal of English and Germanic Philology* 45 (1946): 68–76; Hunter, "Milton's Power of Matter," *Journal of the History of Ideas* 13 (1952): 551–62; John Reesing, "The Materiality of God in Milton's *De Doctrina Christiana*," *Harvard Theological Review* 50 (1957): 159–74; and George Williamson, "Milton and the Mortalist Heresy," *Studies in Philology* 32 (1935): 553–79.

4. Christopher Kendrick offers a Marxian reading of Milton's monism in *Milton: A Study in Ideology and Form* (New York: Methuen, 1986); and Stephen M. Fallon, in *Milton among the Philosophers: Poetry and Materialism in Seventeenth-Century England* (Ithaca: Cornell University Press, 1991), pp. 79–110, argues for the centrality of vitalism to Milton's theology of free will. The most powerful account of Milton's monism is to be found on pp. 193–262 of William Kerrigan, *The Sacred Complex: On the Psychogenesis of "Paradise Lost"* (Cambridge: Harvard University Press, 1983).

inscription of the politically resonant discourses of vitalist science. Chapter 2 allies the action of Marvell's pastoral "Mower" to the action of the "Digger" represented by Marvell's contemporary, the communist visionary Gerrard Winstanley. Chapter 3 investigates the relation of Marvell's images of virginity to the vitalist attestation of virginal power at midcentury; the figure of the virgin surfaces in Marvell's poems to embody a wide range of political, even apocalyptic, forces. The next chapter, on Milton's chaos, maps the Restoration revival of vitalism in *Paradise Lost*. Although Milton had, by the time he published his epic, abandoned the most liberal, egalitarian impulses of his early political writing, his early radicalism surfaces in the poem in the language of a monistic natural philosophy. Chapter 5, on Milton's representation of the expulsion, seeks a new contextualization for the ostensibly authoritative discourse of "eternal providence" in *Paradise Lost*. The tension in Milton's poem between his science of vitalist agency and his seemingly Calvinist providentialism exposes the necessarily dialectical origins of the ideology of liberalism with which Milton has come to be so powerfully associated. The subject of Chapter 6 is the natural philosophy of Margaret Cavendish, the Duchess of Newcastle. In a series of vitalist treatises on the nature of material particles, this Royalist contemporary of Marvell and Milton appropriates the rhetoric of radical Puritan utopianism and opens to explicit view the gendered nature of the poetics of agency and organization that had structured all the period's vitalist texts.

In pursuing the interrelations of religion, science, and politics, I have at many points configured the object of my analysis as a unified discursive field, the set of seventeenth-century linguistic constructions given to the articulation of agency and organization. This discursive field constitutes the mutually informing disciplines of natural and political philosophy, but it also structures and unsettles many of the period's most striking works of literature. The conceptual struggles endemic to revolutionary England were mapped out, no less for Winstanley and Hobbes than for Milton and Marvell, on the terrain of linguistic representation, a terrain whose formal demarcation by strictures of grammatical agency and syntactic organization at once helped and hindered the conceptualization of the agential and organizational problems at hand. My strategy therefore throughout this book has been to test the matter of England's revolutions through an analysis of the poetics of revolution, of the charged rhetoric of action and structure that marks so much of this period's writing. We will see the unacknowledged assumption of the logical interdependence of natural and political philosophy push all the writers studied here into logical traps and literary binds often well beyond the parameters of authorial design. In tracing various entanglements in the frequently unintended literary phenomena of paradox, syntactical confusion, and narrative and thematic contradiction, we are not merely enlivened to the impact of the English and Scientific Revolutions on seventeenth-century letters. This literary perspective can expose as well the fault lines dividing the conflicting motives, interests, and modes of thought that helped impel these revolutions in the first place. It is my hope that this book brings into focus the consequences, at once literary and political, of the alliance, at this moment of England's two great revolutions, of science, poetry, and politics.

JOHN ROGERS

Hatfield, Massachusetts

ACKNOWLEDGMENTS

The research for this book has been supported by many generous sources, including a Whiting Fellowship in the Humanities, a John F. Enders Research Assistance Grant, and an A. Whitney Griswold Faculty Research Fellowship, all administered by Yale University. I am grateful for the assistance afforded by a Short-Term Fellowship at the William Andrews Clark Memorial Library, an American Council of Learned Societies Grant-in-Aid, a W. M. Keck Foundation Fellowship at the Huntington Library, and especially an Eccles Fellowship at the Humanities Center at the University of Utah and a Mellon Fellowship at the Society of Fellows in the Humanities at Columbia University.

I want to acknowledge my appreciation for the encouragement, the scholarly advice, and the diverse intellectual examples provided by John Hollander and John Guillory. I am thankful for the help from Richard Burt, Victoria Kahn, David Quint, Barry Weller, Jonathan Post, and especially Kevin Dunn; each of them read the entire manuscript, at various stages of completion, and offered incisive and fruitful suggestions. Among those who have given useful assistance with the earliest versions of the material here, I must single out George deF. Lord, Leslie Brisman, Claude Rawson, Susanne Wofford, Cristina Malcolmson, William Clark, Jason Rosenblatt, John King, Karen Lawrence, Donald Friedman, Peter Goldstein, Nicholas von Maltzahn, Heather Dubrow, and Barbara Estrin. The members of the Northeast Milton Seminar, the Works in Progress group at Yale, and the participants in a colloquium at the University of Utah Humanities Center challenged me to change and develop chapters 1 and 4. For some timely advice on seventeenth-century English philosophy, I am indebted to Marilyn Pearsall. I wish

to thank Bernhard Kendler, of the Cornell University Press, for his interest
in this project, as well as Teresa Jesionowski and Kim Vivier for the metic-
ulous care they brought to their editing. I will always be grateful, for their
more general gestures of good will and support, expressed in a variety of
ways, to Richard Brodhead, Harold Bloom, David Kastan, Mary Ann Radzinowicz, and Joseph Wittreich. Carol and Gerald Rogers have been steadfast
in their kind support, exemplifying an ideal of the parental encouragement
of the academic.

My deepest thanks I owe to Cornelia Pearsall. Her intellectual guidance
has left its mark on every page of this book, her companionship has sustained
every moment of its composition.

THE MATTER OF REVOLUTION

1 The Power of Matter in the English Revolution

> As for our intellectual concerns, I do with some confidence expect a Revolution, whereby Divinity will be much a Looser, & Reall Philosophy flourish, perhaps beyond men's Hopes.
>
> —ROBERT BOYLE (1651)

The story of the relation among science, politics, and literature in the later seventeenth century begins with a brief but potent burst of intellectual activity at a particular juncture in the Revolution, the interval just before and after the execution of Charles I on January 30, 1649. The period between 1649 and 1652 sees the production of an impressive group of texts: Andrew Marvell's "Horatian Ode" and nearly all the pastoral poems, the communist manifestoes of Gerrard Winstanley, John Milton's *Tenure of Kings and Magistrates*, *Eikonoklastes*, and *First Defence of the English People*, Thomas Hobbes's *Leviathan*, William Harvey's *Disputations on the Circulation of the Blood* and *On the Generation of Animals*, and two volumes of England's first established woman writer, Margaret Cavendish, the Duchess of Newcastle. These particular works, encompassing a wide array of literary genres and discursive modes, engage, quite obviously, a vast and divergent set of interests. But for all their many differences, each of these texts participates in or reacts to one of the least understood intellectual movements in early modern England, a short-lived embrace of philosophical idealism that I identify as the Vitalist Moment. The philosophy of vitalism, known also as animist materialism, holds in its tamest manifestation the inseparability of body and soul and, in its boldest, the infusion of all material substance with the power of reason and self-motion.[1] Energy or spirit, no longer immaterial, is seen

---

1. For discussions of vitalism, see L. Richmond Wheeler, *Vitalism: Its History and Validity* (London: Witherby, 1939), pp. 3–27; Walter Pagel, *William Harvey's Biological Ideas: Selected Aspects and Historical Background* (New York: Hafner, 1967), pp. 251–77; Merchant, *Death of Nature*, pp. 117–26; John Yolton, *Thinking Matter: Materialism in Eighteenth-Century Britain* (Oxford: Oxford University Press, 1984), pp. 1–26; John Henry, "The Matter of Souls: Medical Theory and Theology in Seventeenth-Century England," in *The Medical Revolution of the Seventeenth Century*, ed. Roger French and Andrew Wear (Cambridge: Cambridge University Press, 1989), pp. 87–113; and Fallon, *Milton among the Philosophers*, pp. 79–110.

as immanent within bodily matter, and even nonorganic matter, at least for some vitalists, is thought to contain within it the agents of motion and change. The most shocking exercises in vitalist doctrine fade from respectability by the early 1650s, just a few years into the republican decade. But the discursive excitement generated by the Vitalist Moment does not die after 1652. It is felt still, however nostalgically, in the Restoration science of Margaret Cavendish and, perhaps most powerfully, in Milton's *Paradise Lost*, published in 1667. The science of self-moving matter, for all the participants of the Vitalist Moment, functions as a flexible, politically resonant form of ontological speculation.

The possibility of a conceptual alliance between a politics and a science may not be readily apparent. There was in seventeenth-century England, as, indeed, in other cultures at other times, a large semantic field that contained and combined the language of the otherwise distinct intellectual practices of political and scientific speculation. There stood an assumption, the more unshakable for the infrequency with which it was acknowledged, of a correspondence between the constitution of the physical world and the structure of human society. It was virtually impossible, as Otto Mayr has demonstrated, to discuss the interaction of physical bodies or mechanical parts without importing the vocabulary and values of the world of political relations, or vice versa.[2] It was, specifically, I believe, the twin questions of "agency" and "organization"—two words that establish themselves in the English language at just this moment of the two revolutions—that drew the seventeenth-century discipline of political philosophy into analogical contact with the discipline of science, or what was called in the period natural philosophy. For all their differences, these disciplines shared a primary investment in identifying and understanding the agents of change; their exploration of the latitude and limits of particular types of action—whether the governance of state or the creation of the universe—provoked a corresponding investment in the mapping of the systematic interaction of these agents. We are, no doubt, most familiar with the political and theological manifestation of the period's inquiry into agency and organization. By what agency, human, divine, or otherwise, are the revolutions in ecclesiastical and political government to be effected? Around what power is the family, the nation, or the

---

2. Mayr, *Authority, Liberty, and Automatic Machinery*. I am indebted to Mayr's comprehensive analysis of the analogical relation of competing physical theories to liberal and authoritarian modes of political thought. In a more general consideration of the social function of natural philosophy, Barry Barnes and Steven Shapin have argued that "any perceived pattern or organized system in nature is liable to be employed to express and comment upon social order and social experience," in *Natural Order: Historical Studies of Scientific Culture*, ed. Barnes and Shapin (Beverly Hills, London: Sage, 1979), p. 15.

entire creation to be imagined as organized? To what extent are individuals to be thought the agents of a reorganization of church or state? These were the questions posed endlessly in this age of revolution, not only by political philosophers but by every type of theologian from high Anglican to radical Puritan.

The answers to the problems of action and system were also sought in another camp of midcentury intellectual life, a discursive realm we rarely think to associate with the readily identifiable traditions of political philosophy or Puritan theology: the discipline of science. The same inquiries into cause and structure were pursued with only some rhetorical modification by the period's chemists, alchemists, physicists, and physiologists. By what agency, by either self-impulsion or external force, does a body of matter move? Around what form of power are particles of matter to be imagined as organized? To what extent are material bodies to be thought the agents of their own organization and reorganization? For us, of course, the issue of physical motion bears no necessary ontological weight on the issue of sociopolitical action. But for mid-seventeenth-century English intellectuals, a theoretical statement of agency and organization in one of these disciplines often required (and as often acquired inadvertently), as a metaphysical guarantee of its validity, a thesis of agency and organization from the other discipline. We have in Hobbes perhaps the best example of this fascinating discursive interdependence: the violent collisions of the bodies of matter in Hobbes's natural philosophy bore a crucial homological relation to the nasty and brutish life of men in the competitive, marketplace environment sketched in Hobbes's political philosophy. The scientific figuration of physical motion spoke throughout this period to the nature and scope of human action, while the figurations of system routinely inscribed the contemporary concerns with political order. Both political and natural philosophy functioned as inextricably intertwined literary practices in seventeenth-century intellectual culture, a culture that demanded the construction of theories of agency and organization that could be seen to hold true for all facets of human and natural existence.

In order to shade in the political contours of the mid-seventeenth-century practice of vitalism, it is first necessary to examine the dominant figurations of agency and organization—the very different practices of theological providentialism and scientific mechanism—against which the culture of vitalism defined itself. The image of mid-seventeenth-century intellectual life with which we are most familiar is that of the radical Puritan faith, a derivative largely of Calvinism, in a divine providence guiding not only the lives of individual men and women but controlling the fate of the nation and, ultimately, the finite span of all human history. There was a tendency among Puritans to view all historical progress as the manifestation of God's arbitrary manipulation of forces on earth, a belief in direct divine determinism—a theological position often called "voluntarism"—which the influence of Calvinism supported and strengthened.[3] Midcentury intellectuals, many though not all of them Puritan, employed with increasing vigor a providentialist mode of typological interpretation that found in the welter of current political events the fulfillment of a purposeful and divinely ordained plan.[4]

The faith held by many Puritans in the immediate pull of an arbitrary deity on the motions of daily life was not, however, the age's only expression of determinist agency and centralized organization. Puritan providentialism found a curious and troubled determinist analogue in the new mechanist materialism—the most distinguished theorist of which was Hobbes—which replaced the immediate agency of divine power with the physical impulsion of one body of matter on another. In heading off the possibility of anything like material self-motion, Hobbes insisted, famously, "there can be no cause of motion, except by a body contiguous and moved."[5] For a rigidly mechanistic philosopher such as Hobbes or Descartes, the agent of motion could never be immanent within a material body, which was always to be seen as passive and inert; motive agency was nothing more than the force produced by one body's collision with another, external body, itself passive and inert.[6]

3. Although Calvin himself denounced voluntarism in its strictest theological form, his unrelenting emphasis on divine will situates him, as Dennis Danielson notes in *Milton's Good God: A Study in Literary Theodicy* (Cambridge: Cambridge University Press, 1982), p. 70, as a "voluntarist in practice if not in theory."

4. See the overviews of seventeenth-century Protestant philosophies of history, in Herschel Baker, *The Race of Time: Three Lectures on Renaissance Historiography* (Toronto: University of Toronto Press, 1967); C. A. Patrides, *The Grand Design of God: The Literary Form of the Christian View of History* (London: Routledge and Kegan Paul; Toronto: University of Toronto Press, 1972); Patrides, *The Phoenix and the Ladder: The Rise and Decline of the Christian View of History* (Berkeley: University of California Press, 1964); and G. W. Trompf, *The Idea of Historical Recurrence in Western Thought: From Antiquity to the Reformation* (Berkeley: University of California Press, 1979).

5. *Elements of Philosophy. The First Section, Concerning Body*, in *The English Works of Thomas Hobbes*, ed. William Molesworth, 11 vols. (London: Bohn, 1839), 1:124. Robert Boyle, though opposed to Hobbes on a range of natural philosophical subjects, joined him in maintaining the orthodox position that "motion does not belong essentially to matter," in the *Free Inquiry into the Vulgarly Received Notion of Nature* (written in 1666, published in 1686), in *The Works of the Honourable Robert Boyle*, ed. Thomas Birch, 2d ed., 6 vols. (London: J. and F. Rivington, 1772), 5:210.

6. John Henry, in "Occult Qualities and the Experimental Philosophy: Active Principles in Pre-Newtonian Matter Theory," *History of Science* 24 (1986): 335–81, and Simon Schaffer, in "Godly Men and Mechanical Philosophers: Souls and Spirits in Restoration Natural Philosophy," *Science in Context* 1 (1987): 55–85, both rightly observe that many later-seventeenth-century philosophers, often called "mechanistic," retained a limited vocabulary of occult qualities, active principles, and immaterial spirits. But the materialist philosophy rejected by the monistic vital-

As Hobbes's *Leviathan* makes strikingly clear, this principle of the violent displacement that necessarily accompanies bodily motion was endlessly extendible to the world of human affairs. The strictly mechanical world-view seemed to lend itself to a political philosophy in which power, a power construed implicitly as a physical force, was one of the only admissible determinants of change. The entire mechanistic philosophy, in fact, flourishing for the first time in England in the years after the execution of Charles, offered scientific proof for the necessity and inevitability of a political process of conquest and domination.

It should not be thought that the spiritually minded Puritan rebels, however eager some may have been to conquer and dominate, could embrace a determinist philosophy of agency like Hobbesian mechanism. The idea of a godly reformation held considerable emotional sway throughout this period, and the mechanistic philosophy was seen by many Puritans to result, as if inevitably, in the affirmation of a political power dangerously unaligned with moral goodness or saintly behavior.[7] But although reaction by the pious to the image of a world composed of a mass of spiritless atoms that God had set spinning into motion was often vociferous, the philosophy of mechanism still bore an important subterranean affinity with the determinism of Calvin. Founded on the principle that motion is the result of a body's impulsion by an external power, the mechanical world-view reiterated and subtly reinforced one of the most overwhelming tenets of Calvin's theocentric providentialism: no motion and, therefore, no action of any kind was possible but for the intervention of an outside power. Orthodox Puritans and mechanist philosophers, while writing within distinct, even exclusive, discursive registers, could agree on the impossibility of self-determination in a world dominated by irresistible external forces.

The structural affinities connecting Puritan providentialism and mechanistic materialism ran deeper still. Not only did these two intellectual systems forward an analogously externalist theory of agency, but they came to posit a curiously analogous thesis of centralized universal organization. Although the immediate agent of motion and change was for the mechanists the impulsion of one body by another, the mechanist vision of a world of lifeless atoms logically required the ultimate government of the creation by a fully centralized power, that power invariably identified as an arbitrary, volunta-

---

ists, and the philosophy I denote throughout this book by the term "mechanist," is the uncompromisingly spiritless materialism of Hobbes and Descartes.

7. The best account of the contemporary reactions to the immorality of Hobbesian materialism can be found in Samuel I. Mintz, *The Hunting of Leviathan* (Cambridge: Cambridge University Press, 1962). See also Henry, "Occult Qualities," pp. 352–58.

ristic God. How else, it was asked, by Hobbes and by nearly all the mechanists, was one to imagine the miraculously ordered patterns in which the
lifeless parts of matter found themselves? For Hobbes and for Robert Boyle,
as for many other members of the group of scientists comprising, after the
Restoration, the Royal Society, the inanimate corpuscles of matter constitutive of the physical world were clearly subordinated to and organized, ultimately, by a single force, the Heavenly Father. This arbitrary deity could
impose on nature and enforce at will an arbitrary set of mechanical laws
much as the God of Calvin held arbitrary sway over the lives of his creatures.[8]
As a corollary to their denial of autonomous individual action, Puritan theologians and Anglican scientists forwarded an analogous picture of a world
whose governance lay in the hands of an arbitrary and unrestricted God.

Here we arrive at one of the most awkward conceptual features of mainstream Puritan theology, and one of the most troubling consequences of the
period's homologization of natural and social representation. It is surely a
generally acceptable truth that, in the formulation of one historian of science,
"groups with conflicting social interests developed and sustained . . . different
natural philosophies."[9] But orthodox Calvinist Puritanism, no matter how
staunchly many of its adherents opposed the monarch's arbitrary authority
over the state, possessed at its ontological core a theology of arbitrary rule.
Although the majority of Puritan dissenters were seeking to free church and
property from the bonds of the Stuart monarch, the arbitrary God of their
religious convictions provided, by the simple logic of the organizational imperative, an unwitting figurative sanction for the absolutist, centralized state
championed by their political opponents, the Royalists. The theological nom-
inalism adopted by so many Puritans, as Joan S. Bennett has argued, was in
fact "a predictable and necessary philosophical base for high church Angli-
cans to adopt because of their alliance with the Stuarts."[10] The Royalist
mechanists were in the fortunate discursive position, unlike many Calvinist
Puritans, to exploit this organizational analogy between a centralized nominalist universe and a centralized absolutist state. James R. Jacob and Mar-

8. See Robert Boyle, *The Origins of Forms and Qualities According to the Corpuscular
Philosophy* (1666), in *Selected Philosophical Papers of Robert Boyle*, ed. M. A. Stewart (New
York: Barnes and Noble, 1979), pp. 69–70; and Boyle, *Free Inquiry*, in *Works of Boyle*, 5:192–
210. As late as 1674, in his treatise *The Excellency of Theology, as Compar'd with Natural
Philosophy*, Boyle persisted in subordinating the dynamics of natural process to the arbitrary
deity of his increasingly outdated voluntarist theology. Shapin and Schaffer discuss Boyle's conviction that matter "was devoid of purpose, volition and sentience" and that the "only ultimate
source of agency in the world was God Himself," in *Leviathan and the Air-Pump*, pp. 201–2.

9. Shapin, "Social Uses of Science," p. 101.

10. Joan S. Bennett, *Reviving Liberty: Radical Christian Humanism in Milton's Great Poems*
(Cambridge: Harvard University Press, 1989), p. 11. I am indebted throughout this book to
Bennett's important reassessment, on pp. 6–32, of the alignment of the theological and political
positions held by Milton and his non-Calvinist Puritan contemporaries.

garet C. Jacob have demonstrated convincingly that the commitment to both an arbitrary God and an arbitrary monarch was an important component of the mechanist movement in midcentury England, a movement demonstrably Anglican and Royalist in the political convictions of many of its practition-ers.[11] For all their attempts to establish a disinterested realm of objective knowledge, the members of the Royal Society tended more often than not to envision a corpuscular universe, governed arbitrarily by an absolutist power, that was peculiarly monarchic in design.[12] The structural paradigm that informed their corpuscular science reflected almost exactly the ideal organizational structure of the conservative politics from which it emerged. These philosophers, some of them sponsored after the Restoration by the king, were eager to confer the ultimate authority for motion and change on a single, absolutist source of material organization, stripping at the same time, much like the Calvinists, all power that may have been thought to dwell within matter itself.[13]

11. See James R. Jacob, *Robert Boyle and the English Revolution: A Study in Social and Intellectual Change* (New York: Franklin, 1977); and J. R. Jacob and M. C. Jacob, "Anglican Origins." Related arguments can be found in Brian Easlee, *Witch-Hunting, Magic, and the New Philosophy: An Introduction to Debates of the Scientific Revolution, 1450–1750* (Sussex: Harvester, 1980); Margaret Jacob, *The Radical Enlightenment: Pantheists, Freemasons, and Republicans* (London: Allen and Unwin, 1981); and Francis Oakley, *Omnipotence, Covenant, and Order: An Excursion in the History of Ideas from Abelard to Leibniz* (Ithaca: Cornell University Press, 1984). Jacob and Jacob's useful, if overly intentionalist, politicization of Boyle's voluntarism and his mechanism has troubled some historians of science anxious to preserve the untrammeled intellectualism of Boyle's achievement. But in the critique of their thesis by John Henry, "Occult Qualities," Henry's conjecture that Boyle asserted his belief in voluntarism and the passivity of matter out of "his reluctance to offend the sensibilities of Churchmen" (pp. 356–57) works to confirm rather than deny the argument for the ideological constraints on the period's natural philosophy. Margaret J. Osler, in "The Intellectual Sources of Robert Boyle's Philosophy of Nature: Gassendi's Voluntarism and Boyle's Physico-Theological Project," in *Philosophy, Science, and Religion in England, 1640–1700*, ed. Richard Kroll, Richard Ashcraft, and Perez Zagorin (Cambridge: Cambridge University Press, 1992), also attacks James Jacob's thesis, but she is compelled to concede that Boyle's "providential corpuscularianism may have been utilized, by himself and others, to support a particular ideological position" (pp. 178–79). In writing of the radical sects, Osler avers as well that the Paracelsian belief in the "innate activity of matter and immanentism of this view seemed to provide support for some of the radically democratic ideologies of these groups" (p. 195 n. 37). In the most recent critique of Jacob and Jacob, Malcolm Oster, in "Virtue, Providence, and Political Neutralism: Boyle and Interregnum Politics," in *Robert Boyle Reconsidered*, ed. Michael Hunter (Cambridge: Cambridge University Press, 1994), voices the fear, common to all anti-Jacobian historians of science, that a political reading of Boyle "erodes the epistemological, philosophical, and theological roots of Boyle's philosophy of nature" (p. 32). My own focus on the poetics of organization avoids the vulgarity not only of Jacob and Jacob's exposé of bald political intentions but also of the needless intellectualism of their detractors: to suggest that Robert Boyle was susceptible to the same pressures of the organizational imperative affecting all the period's intellectuals is to accuse the philosopher of neither intellectual dishonesty nor political expediency.

12. See Mayr, *Authority, Liberty, and Automatic Machinery*, pp. 85–92.

13. "At its very origins," M. C. Jacob has argued in *Radical Enlightenment*, the new materialist science "was perceived and used to enhance the power of ruling elites and prevailing

These were the determinist philosophies of agency and organization felt by many to dominate the intellectual landscape in mid-seventeenth-century England. Puritan voluntarism and "Royalist" mechanism, however divergent their constituents, insisted on the truths of external agency and centralized organization. And it is just this insistence on outward force and centrifugal governance—the determinisms represented most pointedly by Calvin and Hobbes—that provided the adversarial background for the emergence of the startling new theorizations of agency and organization of the Vitalist Moment.

The Vitalist Moment

Historians of early modern science have been struck by the intense interest in "chymistry"—a broad term for both vitalist and nonvitalist forms of alchemical speculation—that flourished during the 1650s, a decade in which more alchemical works were published and translated than "in the entire century before 1650."[14] It is true, as J. Andrew Mendelsohn has shown, that a few of the period's "chymical" texts were happily embraced by Royalists, both before and after the Restoration. But it is nonetheless the case, as Mendelsohn has also argued, that there is "dramatic evidence for how quickly after 1649 chymistry became identified with subversion of political and religious order."[15] By all accounts the widespread interest in "chymistry" took its origin in the period's climate of political crisis. It is necessary now to explain the nature of the connections that emerged at the Vitalist Moment to tie a radical science to a radical politics and theology. What are the cultural filiations that enable this identification of philosophical vitalism with religious and political subversiveness?

Christian orthodoxy" (p. 31). Jacob goes perhaps too far in arguing that Boyle segregates motion from matter "*in direct response* to the materialistic pantheism proposed by the radical sectaries, by philosophers such as Winstanley and by political leaders such as Lilburne" (p. 71; emphasis mine).

14. Allen G. Debus, *The Chemical Dream of the Renaissance* (Cambridge: Heffer, 1968), p. 26. See also F. N. L. Poynter, "Nicholas Culpepper and the Paracelsians," in *Science, Medicine, and Society in the Renaissance: Essays in Honor of Walter Pagel*, ed. Allen G. Debus, 2 vols. (New York: Watson, 1972), 1:201–20; and Hugh Trevor-Roper, "The Paracelsian Movement," in his *Renaissance Essays* (London: Secker and Warburg, 1985), pp. 185–95.

15. J. Andrew Mendelsohn, "Alchemy and Politics in England, 1649–1655," *Past and Present* 135 (1992): 34. The burden of Mendelsohn's argument is the "extraordinary fluidity, even inconsistency, in alchemy's ideological alignments" (p. 76). I suspect that Mendelsohn's insistence on the inconsistency of alchemy's political alliances, though a useful corrective, largely reflects his failure to distinguish the dualist mysticism associated with Boehme from the monistic materialism derived from van Helmont.

A few historians have attempted to explain the prevalence of alchemy in the Puritan decade by aligning the progressivist dreams of the Paracelsian alchemists with the more tangible reforms sought by the Puritan rebels; and, indeed, these intersecting groups of intellectuals appear to share a commitment to principles of "reform."[16] But the vague reference to progressivist tendencies is hardly sufficient to justify the unprecedented, and largely unrepeated, burst of interest in the specifically vitalist forms of Paracelsian speculation. The midcentury explosion of alternative materialist philosophies bears a more specific relation than a shared progressive idealism to the contemporary propositions of political reform. It is the period's analogical imperative, I believe, that cultural pressure always pushing for the structural alignment of representations of political and material organization, that best explains this appearance of an alternative science at a moment of political and social conflict. In response to the assumed systemic homologies between nature and polity, a disparate group of mid-seventeenth-century intellectuals, some though not all of whom were on the Puritan left, began to distance themselves from the rhetoric of arbitrary authority at the heart of determinist discourse. The figure of autonomous material agency peculiar to animist materialism provided fruitful conceptual backing for a range of identifiable groups, including politically minded radicals seeking a liberatory conception of individual political agency, and others, motivated less by political than economic concerns, pursuing a principle of free agency in a hypothetically free market. Vitalist agency, charged with the momentum of revolutionary fervor, could soon be invoked for radical cultural ends, by intellectuals as differently motivated as Milton and Margaret Cavendish, as the "power of matter."[17]

When, in *Paradise Lost*, the Father infuses his "vital virtue . . . throughout the fluid mass" of chaos at the Creation (7.234–37), Milton, like a number

---

16. Charles Webster, "English Medical Reformers of the Puritan Revolution: A Background to the 'Society of Chymical Physitians,' " *Ambix* 14 (1967): 16–41; Webster, *The Great Instauration: Science, Medicine, and Reform, 1626–1660* (London: Duckworth, 1975), pp. 27–31, 315–23; P. M. Rattansi, "Paracelsus and the Puritan Revolution," *Ambix* 11 (1963): 24–32; Rattansi, "The Social Interpretation of Science in the Seventeenth Century," in *Science and Society, 1600–1900*, ed. Peter Mathias (Cambridge: Cambridge University Press, 1972), pp. 11–12; Keith Thomas, *Religion and the Decline of Magic* (New York: Scribner's, 1971), pp. 270–71; Christopher Hill, *World Turned Upside Down: Radical Ideas during the English Revolution* (Harmondsworth: Penguin, 1975), pp. 287–305; Hill, *Milton and the English Revolution* (Harmondsworth: Penguin, 1977), p. 329.

17. Milton, *Complete Prose Works of John Milton*, ed. Don M. Wolfe, 8 vols. (New Haven: Yale University Press, 1953–82), 6:322; Margaret Cavendish, "Preface," *Philosophical and Physical Opinions*, 2d ed. (London, 1663), n.p. Kendrick, in *Milton*, has argued for the status of Milton's monistic theory of the "power of matter" as a response to "the commodification of the individual subject's powers by emergent capitalism" (p. 13).

of midcentury vitalists, claims affinity with a philosophy of matter, derived ultimately from the sixteenth-century alchemist Paracelsus, that "designates the unity of matter and spirit as a self-active entity."[18] The theories of the microcosm developed by Paracelsus sketched a map of the creation that linked man and the universe in a self-contained cosmic economy of interflux and exchange. Although Paracelsus himself had viewed the elaborate set of sympathies and correspondences uniting man and the universe as immaterial, his "animistic interpretation of the earth and celestial bodies, and the guiding concept of the *anima mundi*, rendered it easy," according to Charles Webster, "to posit a connection between the physical and organic and psychic world."[19] Through a broadly motivated resuscitation of Paracelsian philosophy in the years of the Vitalist Moment, the old medieval metaphor of the *corpus mysticum*, stripped of its mysticism, was made flesh, expanded into a map of a truly bodily plenum that incorporated all that had been known as psyche and soul into the natural world of moving bodies. In reintroducing figures of reason and sentience into the sphere of material process, the premier vitalist theorists, Jean Baptiste van Helmont, William Harvey, and Francis Glisson, could articulate scientific figurations of agency distinct from the oppressive voluntarism of Calvin on the one hand and, on the other, the amoral ascendancy of spiritless physical force implicit in the materialist philosophies of Hobbes and Descartes.[20] Once the vitalists infused, like Milton's Father, the mass of creation with "vital virtue," the power of matter was in a position to form the theoretical substrate for a revolutionary reconception of power in general.

Many of the vitalists did not stop at the infusion of objects in the natural world with a simple vegetative capacity for movement. For Gerrard Winstanley and Andrew Marvell, the subjects of Chapter 2, the matter of Nature

18. Merchant, *Death of Nature*, p. 117. All quotations from Milton's poetry are drawn from *John Milton: Complete Poems and Major Prose*, ed. Merritt Y. Hughes (Indianapolis: Odyssey, 1957), and will be cited by line number in the text.

19. Charles Webster, *From Paracelsus to Newton: Magic and the Making of Modern Science* (Cambridge: Cambridge University Press, 1980), p. 30. The mid-seventeenth-century interest in the sympathetic connection between the bodily microcosm and the macrocosm is vividly exemplified in John French's 1650 Paracelsian tract *A new light of Alchymie: taken out of the fountaine of nature, and manuall experience* (London: Richard Cotes, 1650), pp. 102–3.

20. I am grouping van Helmont, Harvey, and Glisson here, each of whom propounds a vitalist vision of self-moving matter, despite the institutional and, to some degree, intellectual differences that divided the College of Physicians, of which Harvey and Glisson were members, from the followers of van Helmont. For a discussion of that institutional division as well the intellectual alliance, see Webster, *Great Instauration*, pp. 315–23. The reaction to Helmontian philosophy is studied by Allen G. Debus, "The Chemical Debates of the Seventeenth Century: The Reaction to Robert Fludd and Jean Baptiste van Helmont," in *Reason, Experiment, and Mysticism in the Scientific Revolution*, ed. M. L. Righini Bonelli and William R. Shea (New York: Science History Publications, 1975), pp. 21–47.

is empowered with such foresight that Nature, and not an anthropomorphic God, is figured as pressing man forward to his promised redemption. For William Harvey, as we see in the next section of this chapter, the material substance constitutive of blood is indistinguishable from rational spirit, a force governing bodily processes "with an *eminent providence* and *understanding*, acting in order to a certain end, as if it did exercise a kind of *Ratiocination* or discourse."[21] The vitalist physician Francis Glisson would go even further, suggesting in 1650 that not only the blood but all human flesh should be seen to exercise reason, a reason founded on its tissue-based capacity for sentience and perception.[22] Writing well after the Vitalist Moment, Milton, as we see in Chapters 4 and 5, often in *Paradise Lost* extends Glisson's vitalism from the world of human flesh to the entire organic creation. And Margaret Cavendish, the subject of Chapter 6, goes furthest of all: looking back at the most radical representations of matter at the Vitalist Moment, to the Glissonian tissue of 1650 and the Harveian blood of 1651, Cavendish bestows on even inanimate, inorganic objects all the attributes hitherto reserved for thinking, soulful human beings: "all things, and therefore outward objects as well as sensitive organs, have both Sense and Reason."[23] Her bodies of matter, in possession of these decidedly human faculties, can move only if they choose to move.[24] At a historical moment in which both Calvinist theology and Hobbesian philosophy seemed to lead inexorably to a crippling constraint on human freedom, the vitalist dream of material self-determination could function, as Stephen Fallon has argued, as an ontological justification of the philosophy of free will.[25] But the discourse

21. William Harvey, *De generatione animalium* (1651), quoted here from the English translation, *Anatomical Exercitations, concerning the Generation of Living Creatures* (London: James Young, 1653), p. 454.

22. On the vitalism of Glisson's 1650 treatise *De rachitude*, see Walter Pagel, "Harvey and Glisson on Irritability with a Note on Van Helmont," in his *From Paracelsus to Van Helmont: Studies in Renaissance Medicine and Science*, ed. Marianne Winder (London: Variorum Reprints, 1986), pp. 497–514. Glisson expands the scope of his vitalism from medical hypothesis to full-fledged philosophy in his *Tractatus de natura substantiae energetica s. de vita naturae ejusque tribus primis facultatibus, i. perceptiva, ii. appetitiva, iii. motiva, naturalibus* (London, 1672), the focus of John Henry's study, "Medicine and Pneumatology: Henry More, Richard Baxter, and Francis Glisson's *Treatise on the Energetic Nature of Substance*," *Medical History* 31 (1987): 15–40.

23. *Philosophical Letters: or, Modest Reflections upon some Opinions in Natural Philosophy By the Thrice Noble, Illustrious, and Excellent Princess, The Lady Marchioness of Newcastle* (London, 1664), p. 18.

24. *Observations upon Experimental Philosophy. To Which is added The Description of a New Blazing World* (London, 1666), sig. g2^r.

25. Fallon, *Milton among the Philosophers*, pp. 96–99, 201–2. Fallon argues that after "the publication of *Leviathan* in 1651, [there] raged a debate in which the question of freedom of the will was inseparable from the debate over the nature of substance and the relation of mind and body" (p. 97).

of self-motion could also work to justify the organizational excrescences of free-will theology: the economics of the decentralized distribution of commodities, the politics of popular sovereignty, and the more radical communist politics of egalitarian self-rule.

Natural philosophers of a Royalist bent, including a majority of those who came to be associated with the Royal Society, were more likely to embrace a vision of matter internally devoid of soul, cast about by external forces or immaterial spirits and overseen by a providential God. A more radical science, one we can tentatively associate with the antiauthoritarianism of the sects and the Independents, however, tended to invest all material substance with either spirit or some implicitly psychoid principle.[26] For republican and sectarian dabblers in natural philosophy, such as Milton, or the communist visionary Gerrard Winstanley, matter was endued at the Creation with a divinely sanctioned capacity for self-motion, virtue, and perhaps even reason. Infused, like man, with the "law of nature," rather than forced to obey, like the mechanists' atoms, a raft of mechanical laws arbitrarily established by a voluntarist God, this living matter was entitled, we are almost led to imagine, to exercise its own will freely in the laissez-faire world of creation. The attribution of divine spirit to all the individual bodies and elements in nature functioned, I believe, to guarantee on the level of natural philosophy the possibility of the harmonious interaction among the self-reliant, virtuous, and rational individuals in the decentralized systems of the polity and the marketplace proposed by the period's Independents and radical sectarians. To infuse matter with a rational spirit or motivating force was not only to render unnecessary an omnipotent and directly controlling God. This monistic vitalism could offer as well the evidence, at the very basis of the micro-universe of material parts, for the efficient and harmonious dynamics of an organization—any organization—operating outside the immediate superintendence of a single, centralizing power. Vitalism, in short, banishing the centralizing logics of Calvinism and mechanism alike, secured into the fabric of the physical world a general scheme of individual agency and decentralized organization that we can identify as a protoliberalism. Liberalism per se, as a recognizable set of interdependent ideals structuring a

26. See M. C. Jacob, *Radical Enlightenment*, pp. 65–86. Webster discusses the strong Parliamentary and Puritan ties of the Helmontian and Paracelsian theorists, in *Great Instauration*, pp. 273–82. Refining Webster's thesis, Peter Elmer identifies the specifically sectarian, radical alliances of midcentury Helmontianism and Paracelsism, in "Medicine, Religion, and the Puritan Revolution," in *The Medical Revolution of the Seventeenth Century*, ed. Roger French and Andrew Wear (Cambridge: Cambridge University Press, 1989), pp. 10–45. Henry More, in his 1656 *Enthusiasmus triumphatus*, aligned the religious enthusiasm of the sects with the vitalist belief that "everything has *Sense, Imagination*, and a *fiducial Knowledge* of God in it, *Metalls, Meteors* and *Plants* not excepted" (quoted in Henry, "Occult Qualities," p. 356).

hypothetical society, polity, and economy, would not find an official articulation until later in the century, in the writings of John Locke. But in developing his late, liberal philosophy, Locke would himself look back to the vitalist ontologies of the midcentury radicals.[27] The liberated doctrine of animist materialism, only nominally circumscribed within the sphere of natural philosophy, was emerging at midcentury to map onto the body of creation the abstract principles of moral choice, independent action, and free association that would come to form the cornerstone of early modern liberalism.

The historical analysis of the religious and philosophical underpinnings of English revolutionary sentiment has focused overwhelmingly on the impact of the Calvinist strain of Puritanism. The Calvinist conviction in a course of events determined with care and precision by an arbitrary, voluntarist God is often seen to have provided a potent conceptual foundation for the wide array of Puritan social, religious, and political causes. It was Calvinism, according to Michael Walzer's influential argument in *The Revolution of the Saints*, that enabled Puritans to construct "a theoretical justification for independent political action."[28] Calvinist providentialism permitted Puritan dissenters to imagine their own formidable political agency as an instrument of divine agency, while Calvinist theocentrism structured the politically useful map of nation and cosmos as an organization centered specifically on the arbitrary and powerful God of Hosts. Walzer is no doubt right to claim that Calvinist providentialism provided many midcentury Puritans with a theoretical justification for independent political action. But it is one of the general goals of this book to shift and expand Walzer's nearly orthodox genealogy of the agential and organizational philosophies of the English Revolution. Among the most radical social and political contributions of the Revolution were the calls for decentralization issued by the Levellers in the late 1640s: the drive toward an extension of the franchise, toward the tolerance of the

27. See Yolton, *Thinking Matter*. Yolton explores how Locke's vitalist suggestion, in *Essay concerning Human Understanding*, that "God can, if he pleases, superadd to matter a faculty of thinking" (4.3.6), "raised a storm of protest and discussion right through to the last years of the eighteenth century" (p. 17). Locke had, like Francis Glisson, served as physician to the family of his patron Anthony Ashley Cooper, Lord Shaftesbury, the figure to whom Glisson would dedicate his *Tractatus de natura substantiae energetica* (London, 1672). Locke's early commitment to Helmontian science is charted in Patrick Romanell, *John Locke and Medicine: A New Key to Locke* (Buffalo, New York: Prometheus, 1984), pp. 51–68.

28. Michael Walzer, *The Revolution of the Saints: A Study in the Origins of Radical Politics* (New York: Atheneum, 1976), p. 3. For a critique of Walzer's thesis, see J. G. A. Pocock, *The Machiavellian Moment: Florentine Political Thought and the Atlantic Republican Tradition* (Princeton: Princeton University Press, 1975), pp. 336–39, 374–75; and Bennett, *Reviving Liberty*, pp. 6–32. Another classic statement of the affinity between English Calvinism and Puritan politics is Christopher Hill, "Providence and Oliver Cromwell," in Hill, *God's Englishman: Oliver Cromwell and the English Revolution* (New York: Harper, 1970), pp. 216–50.

religious diversity of the sects, and toward the democratization and decentralization of the guilds. The alienating hierarchalism of predestinarian Calvinism was incapable, I think, of generating or subtending a functionally coherent discourse of nonhierarchical political association; the centralized universe of Calvin could not be brought into a conceptually satisfying structural alignment with the ideals of freedom and decentralization that had begun, in this period, their coalescence into liberalism. The science of vitalism, I believe, exercised its ideological function on a deeper, more ontological level than the Calvinism it opposed, supplying in the form of the science of self-motion a theoretical justification for the more collective mode of political agency and the more inclusive vision of political organization that were among the unquestionable products of the English Revolution.[29]

Many strains of radical sentiment, of course, were never taken up by the Puritan Independents, who had championed the rights of property against traditional authority but had stopped short of accepting the most democratizing proposals of the Levellers or Diggers. None of the writers examined in this book, with the exception of Gerrard Winstanley, imagined that his compelling figurations of a self-active material world would embrace or uphold these most radical of the period's revolutionary desires. Two of them, Harvey and Cavendish, were committed Royalists, before and after the Civil Wars, and could not have displayed less interest in seeing the extension of their "liberal" visions of material organization into corresponding prescriptions for a liberal political state. Despite, however, the ideological differences that separated nearly all the period's vitalists from the thoroughgoing egalitarianism of a Winstanley or an Overton, the organizational rhetoric of vitalism continually pulled even the most conservative vitalists into an unwitting intimacy with the most dangerous figures of the Revolution.

One measure, I believe, of the radical tug of vitalist rhetoric was the effect it had on the representation of sexual hierarchy.[30] The traditional subordination of matter to spirit had always been figured and justified as the reasonable subordination of a female to a male principle of being. And the monistic materialists of the seventeenth century unquestioningly reproduced the inveterate gender signs attached to the categories of matter and spirit.

29. In a similar argument, which looks at the relation of animist materialism to the specific concern of Anglican Church hierarchy, Steven Shapin, in "Social Uses of Science," p. 102, explains that "a spiritually imbued material world provided a usable vision of a self-moving and self-ordering system, independent of superintendence by spiritual intermediaries."

30. The central studies of the relation of philosophies of matter to questions of gender are Merchant, *Death of Nature*, and Evelyn Fox Keller, *Reflections on Gender and Science* (New Haven: Yale University Press, 1985).

In Milton, for example, as W. B. C. Watkins has provocatively observed, "matter is to all intents and purposes the feminine aspect of God."[31] But despite the continued assignment of male and female qualities to spirit and matter, vitalism's general reconfiguration of the relation of spirit to matter compelled, at least rhetorically, a parallel reconfiguration of the relation of male to female. The monist's insistence on the spiritualization of matter worked inevitably to elevate the discursive category of femaleness, traditionally mired in matter, that dualism had helped keep in check. We observe, in the chapters that follow, the egalitarian rhetoric of monistic vitalism draw all our writers to entertain a reconceptualization, if only a provisional discursive one, of the traditionally authoritative hierarchy of the sexes. The egalitarian logic of this vitalist theory seemed implicitly to *necessitate* a feminism, one of the logical discursive consequences of the philosophy of monism that would not be positively embraced or explicitly voiced until the Restoration prose of the monist Margaret Cavendish.

The Royalist scientist Robert Boyle, before his conversion to mechanism, had flirted in his youth with the alchemical vitalism of van Helmont, John Everard, and some of the other midcentury Paracelsians.[32] He had written, at the height of his own excitement with the Vitalist Moment, in 1651, of the exhilarating but dangerous prospects of an intellectual "Revolution, whereby Divinity will be much a Looser, & Reall Philosophy flourish, perhaps beyond men's Hopes."[33] Boyle, I think, was right to anticipate the flourishing of vitalist philosophy at the expense of divinity: in this period's zero-sum economy of energic power, vitalism could only infuse the particles of matter with the vitality and volition it had stripped from the Deity. Boyle was also right, however, to conjecture that the vitalist philosophy might flourish in this period "beyond men's Hopes." The power of matter, in its vitalist formulation, seemed continually to burst the ideological frame fashioned for it by the vitalist philosophers. The diverse vitalist writers studied in this book, not at all identifiable with a single political interest, provide us with examples of the uncontrolled literary effects—textual moments of narrative discontinuity, argumentative contradiction, and rhetorical self-occlusion—wrought by the commitment to a vitalist science, ideologically radical effects that flourish well beyond the recognizable hopes or intentional strategies of their authors.

There lay at the heart of many of the period's expressions of vitalist agency

31. W. B. C. Watkins, *An Anatomy of Milton's Verse* (Baton Rouge: Louisiana State University Press, 1955), p. 63.

32. See J. R. Jacob, *Robert Boyle*, pp. 110–12.

33. British Museum manuscripts, Harley 7003, fol. 180^r, quoted in J. R. Jacob, *Robert Boyle*, p. 97. See the related sentiments in Boyle's *Occasional Reflections* (London, 1665), pp. 13-14.

and organization a discursive logic of egalitarianism so pure and unmodified that its implicit organizational connection to the world of the polity, the marketplace, or the domestic hierarchy of marriage was a matter, at best, of discomfort, at worst, of considerable anxiety. Regardless of the personal commitments of a given vitalist writer, there was, woven into the argumentative fabric of many claims for self-moving matter, an organizational rhetoric so hostile to hierarchy, any hierarchy, that nearly all its adherents, at one point or another, were compelled to retreat from its broadest social and political implications. The clearest and most paradigmatic theorist of vitalist agency and organization, William Harvey, is also that figure whose writing is most fissured by its ironic organizational relation to his politics. I propose that we turn, by way of introduction, to a consideration of the crisis of agency and organization that follows the enunciation of the power of matter in Harvey's physiological treatises. Unlike Marvell, Milton, or Cavendish, each of whom articulates a monism within a more or less self-consciously literary frame, Harvey is committed quite obviously to a nonliterary elaboration of a single scientific truth. The manifestly nonfictive status of his texts provides us with a useful model for the way in which the figurative and analytic crises generated by vitalism in all later-seventeenth-century writing can escape the realm of authorial intention and express crises endemic to the intellectual culture at large. The century's most instructive instance of the revolutionary impact of vitalism can be found in the century's most celebrated account of a literal revolution, Harvey's thesis of the circulation of the blood. Harvey's science of circulation introduces for us the antithetical paradigms of agency and organization that structure not only the natural philosophy of the revolutionary period but also the less theoretical, far more deliberately fictional texts of the Marvellian lyric and the Miltonic epic.

William Harvey and the Revolution of Blood

In a preface to a translation of the works of William Harvey, Zachariah Wood, in 1653, addresses the celebrated doctor as the "seditious Citizen of the Physicall Common-Wealth!"[34] Wood's image of sedition, of course, points to the conceptual revolution behind Harvey's discovery of the blood's circulation, an act of rebellion against contemporary medical orthodoxy that easily qualified Harvey as the century's premier "disturber of the quiet of

---

34. William Harvey, *The Anatomical Exercises of Dr. William Harvey . . . with the Preface of Zachariah Wood, Physician of Rotterdam* (London, 1653), sig. *4ʳ.

Physicians!"[35] But Wood's epithet speaks to more than Harvey's role as a revolutionary physiologist disturbing the intellectual commonwealth of seventeenth-century Galenic practitioners. The "Physicall Common-Wealth" that Harvey engaged in his natural philosophy was first and foremost the corporeal commonwealth of the human body. Throughout his long career as physician and natural philosopher, Harvey conceived of the body, with all its mechanisms for governance and control, with its drive to maintain the stability and health of its members, as a polity. As a student of the "Physicall Common-Wealth" Harvey had already embedded a figurative revolution within his two major accounts of the blood's circulatory motion. As Christopher Hill has persuasively demonstrated, the figurative outlines of Harvey's circulatory theories bear an uncanny resemblance to the figurative outlines of the period's political philosophies.[36] For Harvey, the blood begins to assume the same status within the body that Harvey himself enjoyed within the medical community: by 1649, in the ferment of the Vitalist Moment, he casts the blood as the "seditious Citizen of the Physicall Common-Wealth."

Throughout his career Harvey fashioned his explanations of the purpose of the circulation of the blood within the literary parameters of political philosophy. It is as an account of the ideal political body, a genre we recognize as utopia, that we must examine Harvey's most important contributions to medical science. The treatises on the circulation of the blood have been typically dissected by historians of science for an understanding of the experimental process by which Harvey discovered the now recognized physical facts concerning the blood's motion.[37] But the intellectual challenge for Harvey himself was not so much the fact of circulation: this was a point about which he appears to have experienced little doubt. Harvey was concerned instead with what he imagined to be the agency, or force, that lay

35. Wood, "Prefatory Epistle," *The Anatomical Exercises of Dr. William Harvey*, sig. *4^r.

36. Christopher Hill, "William Harvey and the Idea of Monarchy," *Past and Present* 27 (1964); reprinted in *The Intellectual Revolution of the Seventeenth Century*, ed. Charles Webster (London: Routledge, 1974), pp. 160–81. Responding to Hill in "William Harvey: A Royalist and No Parliamentarian," *Past and Present* 30 (1965), reprinted in *Intellectual Revolution*, pp. 182–88, Gweneth Whitteridge counters Hill's politicization of the inconsistencies in Harvey. She continues the critique in her book, *William Harvey and the Circulation of the Blood* (London: Macdonald, 1971), pp. 215–35, denying "that Harvey's views on the relationship of heart and blood varied to any considerable extent throughout his life" (p. 232). Concurring with Whitteridge, Pagel, in *Harvey's Biological Ideas*, p. 341, insists that the discrepancies in Harvey's theories can be ascribed simply to "a scientific—observational—motive." In *Flesh and Stone: The Body and the City in Western Civilization* (New York: Norton, 1994), published after I wrote this chapter, Richard Sennett, following Hill, meditates on the impact of "Harvey's discoveries about healthy circulation in the body" on the "new capitalist beliefs about individual movement in society" (p. 256). I am grateful to Timothy Raylor for bringing the Hill–Whitteridge debate to my attention.

37. See, for example, Whitteridge, *William Harvey*.

behind the blood's circular motion.[38] In his lifelong meditation on the question of agency and the parallel question of bodily organization, Harvey presents himself as an exemplary figure for this study of the conflictive discursive structures of revolutionary England.

Having established as early as 1616 the empirical phenomenon of the self-enclosed circuit of moving blood, Harvey devoted his theoretical energy to the determination of "whether the blood be mov'd or driven [by the heart], or move it self by its own intrinsecall nature."[39] The question of the precise location of what he called the "pulsifick force" (p. 150), a force that he imagined had to reside either in the heart or in the blood itself, presented Harvey with a problem that occupied him throughout his career. From 1616 to 1649, he arrived at a number of formulations for two distinct, and irreconcilable, answers to the question of the ultimate source of the blood's impulsion. His articulation of the mechanisms attending the powers behind bodily process drew him, as if inescapably, to diagram distinct and irreconcilable models of the body politic. Harvey's conflicting answers to the question of sanguineous agency fall in line with the two organizational discourses—authoritarian and liberal—that would come to dominate the general consideration of causation and structure for the remainder of the century.

In his pioneering *De motu cordis* of 1628, the world of the body Harvey explores is a kingdom.[40] The circulatory system is structured, like the solar system, around a centralized and implicitly monarchic power. In a rather effusive dedication to Charles I, Harvey explains, "*The Heart* of creatures is the foundation of life, the Prince of all, the Sun of their Microcosm, on which all vegetation does depend, from whence all vigor and strength does flow. Likewise the King is the foundation of his Kingdoms, the Sun of his Microcosm, the *Heart* of his Common-Wealth, from whence all power and mercy proceeds" (p. vii). Harvey's felicitous analogy between heart and king, however, quickly establishes itself as more substantial and less self-consciously literary than we might at first expect. The figure of the monarchy of the heart, spilling out of Harvey's dedication to the king, is central to the main text's formal articulation of its thesis. In the body of *De motu*, we learn that the heart is the "Prince in the Commonwealth, in whose person is the

38. See John G. Curtis, *Harvey's Views on the Use of the Circulation of the Blood* (New York: Columbia University Press, 1915).

39. William Harvey, *The Anatomical Exercises of Dr. William Harvey, De Motu Cordis 1628: De Circulatione Sanguinis 1649: The First English Text of 1653*, ed. Geoffrey Keynes (London: Nonesuch, 1928), p. 164; all citations of *De motu* (*On the Movement of the Heart*) and *De circulatione sanguinis* (*On the Circulation of the Blood*) are taken from this edition and cited by page number in the text. For a discussion of Harvey's early musings on the "primacy of the Heart" and the "antiquity of the blood," see Whitteridge, *William Harvey*, pp. 215–16.

40. See Hill, "Harvey and the Idea of Monarchy."

first and highest government every where; from which as from the original and foundation, all power in the animal is deriv'd, and doth depend" (p. 115).

Like any number of the political philosophers with whom he was contemporary, Harvey was obliged in his study of the body to address the question of the control over finite resources, the most important resource being, of course, nutritional. Not only is Harvey's absolutist prince—the heart—fully in control of the motions of sanguineous circulation, but he "doth his duty to the whole body, by nourishing, cherishing, and vegetating" the "outward parts," those bodily extremities that function, Harvey tells us, as the heart's "dependents" (pp. 59–60). However absolutist the structure of the "principality of the heart," this prince can be said to rule his dependents with a kind and gentle paternalism: with a firm control over what Harvey calls the "oeconomy of the body," the heart opens itself up as a storehouse of corporeal nutrition. In a passage from the 1628 *De motu*, Harvey explains that the heart "alone of all parts . . . does contain in its concavities, as in cisterns, or a celler (to wit, ears or ventricles), blood for the publick use of the body" (p. 95). He acknowledges that the heart, like all other organs, has blood "for its private use": the heart's "coronal vein and arterie" exist for the sole purpose of funneling nutrition back to itself. But the heart, alone of all the organs, has an additional, "publick" function, which is to supply nourishment to all the private subjects of the body's kingdom. The allocation of this nourishment to the bodily dependents, or extremities, is, of course, the sole prerogative of the heart: "the heart only is so plac'd and appointed, that from thence by its pulse it may equally distribute and dispence . . . to those which want, and deal it after this manner, as out of a treasure and fountain" (p. 95). Through the heart's benevolent though mercantilist manipulation of the process of systole, every want of the bodily extremities is met.

Harvey was thus able to account for the heart's role in the circulation of the blood by recourse to a process that was by almost any account at the center of seventeenth-century biological theory: the process of digestion. It was the science of digestion that availed itself, more than any other process, to the metaphysical justification of political governments, since, presumably, the control over the distribution of food is one of a government's most important sources of power. Dr. Walter Charleton was able to write in his 1659 treatise on digestion that "the most perfect Model or Form of Government . . . is the Body of Man."[41] The Royalist Dr. Charleton was particularly canny, however, in refusing to identify, in the troubled days of the English republic,

41. Walter Charleton, *Natural History of Nutrition, Life, and Voluntary Motion* (London, 1659), sig. A3$^\mathrm{v}$.

precisely what model or form of government the body of man assumed. When Harvey publishes, in 1649, his next major treatise, *De circulatione sanguinis*, it is the problem of the body's "model or form of government" with which he is most consumed. And it is once again the question of agency, or the source of circulatory impulsion, on which he focuses his attention. But although Harvey does not retract or alter the essential description of circulation he had offered twenty years earlier, he performs an acrobatic theoretical reversal that is in many ways as surprising and unaccountable as his initial discovery. By 1649, Harvey appears to have applied to his understanding of the human body the principal tenets of vitalism. Blood was no longer that fluid simply pumped and circulated by the heart; empowered by a vitalist infusion of spirit and energy, Harveian blood is now in possession of its own "native heat, call'd innate warmth" (p. 188). The consequence of this vitalist turn is extraordinary. The attribution of vitalist agency to the blood necessitates a reconfiguration of the entire map of bodily organization. The circulation of the blood, in 1649, is no longer effected by "the heart, but by the meer impulsion of the blood" (p. 183), which has supplanted the heart as "the first efficient cause of the pulse, as likewise to be the common instrument of all operations" (p. 188).

By means of a systematic displacement of all the attributes of power from the heart onto the blood, Harvey subjects the government of the circulatory system to a radical revolution. Whereas in 1628 the blood is said to "return to the heart, as to the fountain or dwelling-house of the body" (pp. 59–60), after 1649 it is the blood itself that is the "fountain of Life."[42] In 1628 the heart "doth his duty to the whole body" as a "familiar household-god" (p. 59), but in the revised circulatory schema it is the blood that "*like a Tutelar Deity, is the very soul in the body.*"[43] It is now the liberally disseminated powers intrinsic to the blood itself that oversee the trade of heat and food through artery and vein. The heart in 1649 is still, as in 1628, the cistern or cellar of bodily nutrition: "The heart is to be thought the Ware-house . . . of the blood" (p. 188). But no longer the "prince" of the corporeal commonwealth, the heart has no control over the distribution and allotment of bodily nutrition. A mere receptacle for the blood, the heart literally has no more agency, or capacity for action, than an actual warehouse: the heart is quite simply "made to be serviceable" to the blood, having been "erected for the

---

42. Harvey, *Anatomical Exercitations*, p. 278. (See note 21.) In the *Anatomical Lectures* (1616), Harvey had written, following Aristotle, "wherefore seeing that the heart imparts heat to all the parts and receives it from none, it is the citadel and abode of heat, the presiding god of this edifice, the fountain and conduit-head" (quoted in Whitteridge, *William Harvey*, p. 220).

43. Harvey, *Anatomical Exercitations*, p. 283. The blood is also described as a "domestick houshold-God" on p. 273.

transmission, and distribution" of it.[44] So removed from its position of chief agent and organizing center, the heart is impelled by the blood, the other organs, and the bodily extremities to continue its activity of pumping. If the heart through its pumping can be said to govern the body at all, it is without question a government with the consent of the governed. Demoted unequivocally from its former status as privileged member, the heart, primus inter pares, is no longer superior in kind, Harvey tells us, to the spleen or the lungs. In Harvey's radical recharting of bodily order, we see the origins of the word new to later-seventeenth-century English, *organization*, a term emerging directly from the discourses of vitalism which could describe at its inception the theoretical determination of the priority or equality of bodily organs.[45]

In the later, 1649 tract on the circulation of the blood, published shortly after the execution of King Charles I, the blood has clearly usurped the heart as the center of the bodily polis and as the distributor of bodily nourishment. Christopher Hill, the one historian who has addressed the unmistakable political resonances of the theoretical inconsistencies in Harvey's science, suggests that Harvey, as if in a bid for patronage, must have deliberately cast his theory in a politicized language to flatter the given moment's political leadership. In a monarchy, Hill implies, the heart is most conveniently imagined a king, and in a republic, it is more prudent to demote the proud heart to a simpler functionary of the more powerful blood. It is true that seventeenth-century political theorists had employed the image of the heart's sovereignty over the body with the express purpose of justifying the king's arbitrary power.[46] But surely it is simplistic to assume that Harvey's complex

---

44. Harvey, *Anatomical Exercitations*, p. 274. See Whitteridge, *William Harvey*, pp. 223, 228.

45. The first seventeenth-century use of "organization" cited in the *OED* is the vitalist Henry Power's discussion, in his *Experimental Philosophy*, 3 vols. (London, 1664), of the "Organization of the Body" (1:82). But this word appears earlier in the period, specifically in the context of the dualist critique of the vitalist reorganization of the human body. Henry More, for example, rejects the suggestion of those vitalists, "which are over-credulous concerning *the powers of the Body*, that *Organization* may doe strange feats," in *The Immortality of the Soul* (London, 1659), cited here from More, *A Collection of Several Philosophical Writings* (London, 1662), p. 77. Ralph Cudworth would later, in *The True Intellectual System of the Universe* (London, 1678), dismiss those "who will attribute Life, Sense, Cogitation, Consciousness and Self-enjoyment . . . to Blood and Brains, or mere Organized Bodies in Brutes."

46. Affirming the monarch's arbitrary right to give preference to the English aristocracy, Edward Forset writes, in *Comparative Discourse of the Bodies Natural and Politique* (1606), "the heart though it spreadeth his arteries all over the bodie, yet hee beateth and worketh more strongly with his pulses in one place than in another. . . . Why then should it be grudged at, if the nobilitie and gentry of the land . . . be better stored and furnished than the meaner of the people?" (p. 45; quoted in Annabel Patterson, *Fables of Power: Aesopian Writing and Political History* [Durham: Duke University Press, 1991], p. 118).

meditations throughout his career on the problem of agency and the origins of bodily power served no purpose but to elevate himself by cajoling king or parliament. We should be able, I believe, to formulate a historical understanding of Harvey's representations of the relationship of heart to blood without resorting to unreflective political intentionalism. Although Harvey's medical treatises clearly expose his participation in the politically resonant discursive culture of his age, they do not therefore necessarily qualify as carefully encoded allegories of midcentury politics. We can, I think, arrive at a more flexible understanding of the discursive associations that tie a scientific treatise, or any representation of agency and organization, to contemporary political conflict.

As we have seen, it was in the late 1640s that nonauthoritarian philosophies of organization made themselves available to the analysis of cause and system across the spectrum of the disciplines. Other physiologists, such as Jean Baptiste van Helmont and Francis Glisson, began, like Harvey, to ascribe bodily processes not to centralized powers, such as the soul, the heart, or the brain, but to local and metabolic forces.[47] Their decentralized paradigms of bodily organization had simply been unimaginable before the mid-seventeenth century. And I would argue that these new maps of physiological order constituted in some way a curious engagement of the first and most influential model of decentralized organization: the economic paradigm of the self-regulating market that had been theorized for the first time in the 1620s to promote a nearly laissez-faire program of foreign trade.[48] The founders of this early avatar of free-trade economics, Thomas Mun and Edward Misselden, had conjectured that goods circulate most freely and the marketplace is most harmonious in a market exempt from the intrusive practices of monarchic price fixing and the granting of monopolies.[49] The sovereign's control over currency and trade, they argued, could be replaced with the predictable operation of the autonomous laws of the market. Their utopian proposals for a self-regulating market for the exchange of goods would find a curious reflex in the work of many of the intellectuals of the Vitalist Moment: the anticensorship model of the free, unlicensed flow of information, the radical political model of a popular sovereignty in a newly decentralized

47. See Jean Baptiste van Helmont, *Ortus Medicinae* (London, 1648); translated into English as *Oriatrike, or Physick Refined* (London, 1662); and Francis Glisson, *De rachitude* (London, 1650); translated into English as *A Treatise of the Rickets* (London, 1651).

48. Joyce Oldham Appleby, *Economic Thought and Ideology in Seventeenth-Century England* (Princeton: Princeton University Press, 1978), pp. 3–51.

49. See Thomas Mun, *A discourse of trade, from England unto the East Indies* (London, 1621); and Edward Misselden, *Free Trade. or, the meanes to make trade florish* (London, 1622) and *The Circle of Commerce. Or the Balance of Trade* (London, 1623).

state, and the Arminian theological model of a world released from the interventionist behavior of the arbitrary and whimsical God of Calvin.[50]

Incapable at its mid-seventeenth-century inception of confining itself to the restricted domains of the polity and the marketplace, the discourse of liberalism spread to the farthest reaches of cultural expression: like the spirited blood that for Harvey circulated throughout the entire human body, early liberalism labored to distribute itself throughout the whole body of contemporary intellectual practice. With the emergence, in the years of the Civil Wars and Interregnum, of the study of the independent "laws" of both the physics of creation and the economics of the marketplace, there developed a discourse of nonintervention that figured the orderly interaction of a multiplicity of individual agents governed not by a solitary sovereign but by what Harvey called in 1649 the "regulating of Nature, an internal principle" (p. 187). This widely applicable liberalism, one of the first discursive manifestations of that process Max Weber named the rationalization of society, sought to subject all figures of centralized agency, God as well as king, to the "Primitive rules," in the phrase of the Cambridge Platonist John Smith, around 1650, "of God's Oeconomy in the World."[51]

This is the broad discursive context in which Harvey came to embrace a vitalist model of agency and a liberal model of organization. It is one of this period's many ironies that the Royalist Harvey's science found a direct and unabashed supporter in James Harrington, who cited Harvey's *Circulatio* repeatedly in justifying the organization of his republican polis in his political utopia of 1656, *The Commonwealth of Oceana*: the senators of Oceana's parliament rotate, Harrington explained, just as the blood circulates through the human body.[52] Harrington took from Harvey the formulation of a political philosophy founded on a principle of bodily government: "Certain it is, that the delivery of a Model of Government (which either must be of no effect, or imbrace all those Muscles, Nerves, Arteries and Bones, which are necessary to any Function of a well-order'd Commonwealth) is no less than political Anatomy."[53] The relation of pure physiological anatomy to the in-

50. For an analysis of the midcentury emergence of a free-market rhetoric of information, see Kevin Dunn, "Milton among the Monopolists: *Areopagitica*, Intellectual Property, and the Hartlib Circle," in *Samuel Hartlib and Universal Reformation: Studies in Intellectual Communication*, ed. Mark Greengrass, Michael Leslie, and Timothy Raylor (Cambridge: Cambridge University Press, 1994), pp. 177–92.

51. John Smith, *Select Discourses* (London, 1660 [published posthumously]), p. 154.

52. James Harrington, *The Political Works of James Harrington*, ed. J. G. A. Pocock (Cambridge: Cambridge University Press, 1977), pp. 248, 287. See I. Bernard Cohen, "Harrington and Harvey: A Theory of the State Based on the New Physiology," *Journal of the History of Ideas* 55 (1994): 187–210.

53. Harrington, *The Art of Law-Giving* (1659), in *Political Works*, p. 656.

tellectual discipline Harrington called "political Anatomy" was not so much one of literary analogy as it was a culturally inescapable homology. Harvey's deployment of a liberal discourse of political organization, one easily reappropriated by a liberal politician such as Harrington, has more to do with the history of compelling patterns of discursive explanation than with literary strategies cunningly deployed for political gain. There seems every reason to believe that in the late 1640s the acceptance of a monistic image of self-moving, spiritualized blood pressed Harvey into a theorization of bodily organization that we can identify loosely as a liberal republicanism.

We have seen how Harvey engaged the new logic of the self-sufficient system operating solely by the laws of self-interest and self-determination. I want now, however, to shift the focus of this investigation and address the possibility that the particular shape and thematic contour Harvey seems to have lent his accounts of circulation derive from an older, less theoretical expression of organizational philosophy. I refer to the classical analogy between the state and the body that, according to Leonard Barkan, "was already a commonplace in Plato's time not only among political philosophers using anatomical description but also among physicians describing anatomy in social or political terms."[54] Harvey's politicized representations of the human body, I believe, bespeak an engagement throughout his career of the analogical fable of the Belly and the members, a fable—at least as old as Plutarch and Livy—retold in England by Camden, Spenser, and Sidney but recounted most famously, of course, by the character Menenius Agrippa in Shakespeare's *Coriolanus* (1608).[55] In Shakespeare, as in Plutarch, the occasion for the fable arises when the hungry citizens of Rome begin conspiring to rebel against the nobility, whom they accuse of hoarding the grain the citizens themselves have labored to produce. The wise and aristocratic Menenius tells the citizens a story that appears to quiet them: a fable about a body in which the "members," or the bodily extremities, rebelled against the Belly. The politically structured body sketched by Menenius has as its center

54. Leonard Barkan, *Nature's Work of Art: The Human Body as Image of the World* (New Haven: Yale University Press, 1975), p. 65.

55. The sixteenth-century philosopher Paracelsus, like many later early modern theorists of digestion, also relies on the terms of the Belly fable, examining the stomach's "work" on behalf of the body's commonweal; see Walter Pagel, *Paracelsus: An Introduction to Philosophical Medicine in the Era of the Renaissance*, 2d ed. (Basel: Karger, 1982), p. 155. In an article that has influenced my own understanding of the liberal figurations of circulation in this period, Dunn examines the role of Menenius's fable in William Potter's *The Key of Wealth* (1650), in "Milton among the Monopolists," pp. 183–86. Menenius's fable is invoked as well by Robert Boyle in a discussion of the interaction of the elements, in the 1661 *The Sceptical Chymist* (New York: Dutton, n.d.), p. 106. Patterson discusses the political uses of the fable of the Belly and the members in *Fables of Power*, pp. 11–38.

the cellar or warehouse of nourishment on which the peripheral bodily members depend. But the warehouse of food, for Menenius, of course, is not the heart but the Belly, who is said to have reminded his mutinous dependents that "I am the store-house and the shop / Of the whole body" (I.i.133–34).[56] The imperious Belly, which Shakespeare's Menenius goes on shortly to identify as the aristocracy, assumes that responsibility of distributional center that will define the purpose and obligation of the Harveian heart in 1628: the Belly dissuades the bodily extremities from rebelling, by reminding them of their dependence on him for food, for

> I send it through the rivers of your blood,
> Even to the court, the heart, to th' seat o' th' brain;
> And, through the cranks and offices of man,
> The strongest nerves and small inferior veins
> From me receive that natural competency
> Whereby they live.
>
> (I.i.135–40)

There is a logical problem at the heart of the fable of the Belly which Menenius himself attempts to evade but which Shakespeare's scene works subtly to disclose. The Belly is confronted with the problem of productive agency. Although he would like to claim himself the productive origin, as well as the governor, of "that natural competency / Whereby they live," the Belly must reluctantly acknowledge that he, too, is a dependent, in the position of receiving the general food from a multiplicity of self-active agents: "True it is, that I receive the general food at first / Which you do live upon" (I.i.130–32). As a storehouse of bodily nourishment, the Belly must still obtain that food from somewhere, and although Menenius never names the ultimate source, it is naturally the laboring bodily extremities that fulfill the role of supplier. The function of the fable of the Belly, for nearly all its early modern expositors, was to elide the awkward and often unacknowledged distinction on which all authoritarian societies were founded, the separation between the public center of power and governance and the private agents of production. I believe it is

---

56. *The Riverside Shakespeare* (Boston: Houghton Mifflin, 1974), p. 1397. Subsequent citations of *Coriolanus* are taken from this edition and cited by act, scene, and line number in the text. A loose identification of heart and belly is already implied by the semantic vagueness of the seventeenth-century "belly." Robert Burton, for example, writes in *The Anatomy of Melancholy*, ed. Holbrook Jackson, 3 vols. (London: Dent, 1932), of "the chest, or middle belly, in which the heart as king keeps his court, and by his arteries communicates life to the whole body" (1:150). Robert A. Erickson cites this passage in "William Harvey's *De motu cordis* and 'The Republic of Letters,' " in *Literature and Medicine during the Eighteenth Century*, ed. Marie Mulvey Roberts and Roy Porter (London: Routledge, 1993), p. 60.

possible to read the utopian body fable structuring Harvey's final treatises on
the circulation as a discursive attempt to reintroduce the organizational distinction so cleverly hidden in Menenius's tale.

Coriolanus's notorious conservatism involves a failure to articulate fully
even the possibility of a system of organization other than the rigid oligarchy of
the Roman senate: "To curb the will of the nobility," Coriolanus proclaims, as
a spokesman for the play's powerfully authoritarian organizational philosophy,
is to "live with such as cannot rule, / Nor ever will be ruled" (III.i.39–41). Between 1628 and 1649 Harvey arrived at a science of systemic distribution that
found a perfectly reasonable order of rule outside the conventionally figured
hierarchy of willful organizing center and unruly, laboring periphery. His de-
centralizing pantheism of the blood would surely have found acceptance by his
far more radical contemporaries, the Digger Gerrard Winstanley and the Leveller Richard Overton, men for whom a monistic natural philosophy and a republican (or even egalitarian) politics were simply inextricable.[57] In this
deflation of the centralized system on behalf of nutritional equity, Harvey pre-
dates by three years Winstanley, England's first communist visionary, who established in his manifesto of 1652 a system of the communal warehouse
distribution of food that resembles nothing so much as Harvey's 1649 account
of the heart's forced warehousing of communal blood: "As every one works to
advance the Common Stock," explained Winstanley, "so every one shall have
a free use of any commodity in the Store-house, for his pleasure and comfortable livelihood, without buying and selling, or restraint from any. . . . And
these general Store-houses shall be filled and preserved by the common labour
and assistance of every Family."[58] The storehouse, as both Harvey and Winstanley argued in implicit disagreement with Menenius, is not the source of
the food it stores; it is simply, like Harvey's dethroned heart, "made to be serviceable" to the real agent or agents of production.[59] The true agents of economic circulation, for these vitalists, could organize themselves in orderly and
predictable fashion "without," in Winstanley's words, "restraint from any." Far
from dissolving into chaos, the liberally disseminated agents of Winstanley's
egalitarian utopia would, like Harvey's blood, behave as harmoniously "as now
. . . under kingly government."[60]

---

57. See Richard Overton's monistic, mortalist treatise of 1644, *Mans Mortalitie*, ed. Harold
Fisch (Liverpool: Liverpool University Press, 1968). David Mulder studies Winstanley's ties to
the alchemical ferment of the mid-seventeenth century in *The Alchemy of Revolution: Gerrard
Winstanley's Occultism and Seventeenth-Century English Communism* (New York: Lang, 1990).

58. Gerrard Winstanley, *The Law of Freedom*, in *The Works of Gerrard Winstanley, With
an Appendix of Documents Relating to the Digger Movement*, ed. George H. Sabine (Ithaca:
Cornell University Press, 1941), p. 583.

59. Harvey, *Anatomical Exercitations*, p. 274.

60. *Works of Winstanley*, p. 584.

I do not want to suggest that Harvey would have invited or acknowledged the intimate discursive allies he had in his Digger contemporaries. But to expose the ties that bind a theoretical vitalist like Harvey to a political radical like Winstanley is merely to state how easily the philosophy of monistic vitalism could become enwrapped in the strands of revolutionary political sentiment. The larger field of alchemically oriented materialism, of which Harvey's late physiology is simply a subset, was seen by many at the Vitalist Moment to provide a scientific foundation for the moral and political reforms sought by revolutionary Nonconformists. In the strictly political sphere, Harvey appears to have supported the king quite as much as his post as Royal Physician demanded; we know he complained bitterly of his loss of some valuable notes and papers at the hands of Parliamentary troops who "rifled his lodgings in Whitehall early in the Civil War."[61] But the conceptual foundation of Harvey's 1649 model of bodily organization is nonetheless undeniably antimonarchic. The lineaments of a radical egalitarianism identifiable in Harvey's late descriptions of the body attest to the overwhelming power, in the England of the late 1640s and early 1650s, of the newly available theories—many of them economic rather than political in origin—of nonauthoritarian agency and organization. Harvey, whose father and four of whose brothers were successful merchants, was, perhaps like many Royalist capitalists in this period, in the discursively awkward position of propounding both a decentralized economy and a centralized political state.[62] It is a testament, I believe, to the discursive power of the "power of matter" that the "political Anatomy" of the king's own physician, committed fully to the authoritarian polity, could be absorbed into the dangerously antiauthoritarian logic of the Vitalist Moment.

A Difficult Birth

The establishment of the new liberal theory of agency and organization was not, for any of the participants in the Vitalist Moment, either immediate or painless. For each of the writers examined in this study, the construction of a liberal discourse reveals the same signs of struggle laid bare in the

61. Whitteridge, *William Harvey*, p. 82; see also Whitteridge, "William Harvey," pp. 187–88; and Pagel, *Harvey's Biological Ideas*, pp. 343–44.

62. I am indebted to Kevin Dunn, who brought to my attention the mercantile background of the Harvey family and the extraordinary fact that one of the early theorizations of a market economy, Lewes Roberts's *Marchants Mappe of Commerce* (1638), was dedicated to seven Harveys, including "Wm Harvey D. of Phys." The mercantile strength of the Harvey family is discussed in Geoffrey Keynes, *The Life of William Harvey* (Oxford: Clarendon, 1966), pp. 128–33.

treatises of Harvey: the assertion, even the theoretical assertion, of noncentralized agency assumes the usurpation of that authority previously imagined
as central. This act of usurpation draws invariably in its wake a discernible
crisis of authority, a crisis that it is the work of many of the period's literary
texts, of whatever political orientation, to engage and, in some cases, to attempt to resolve. The organizational implications of vitalism can be so far-
reaching, the logical extension of monism so far to the left of the ideological
commitments of its proponents, that one of vitalism's most fascinating literary effects is the extraordinary discursive bind in which it appears to trap
its adherents. We examine in the next chapter, for instance, the awkward
affinity that Marvell's monistic poetics forms with the rhapsodic utopianism
of Gerrard Winstanley. We look in Chapter 4 at the literary crisis vitalism
precipitates in *Paradise Lost*: much of the animist materialism Milton espoused in his *Christian Doctrine* must be transferred in his epic, for specifically political reasons, to the discredited voice of Satan. For an introduction,
however, to the crises of authority with which all the writers discussed in
this book were confronted, I propose we turn, once again, to Harvey. I want
to conclude with an analysis of a critical moment in Harvey's late work on
sexual reproduction, *De generatione animalium*. Here we have, I believe, the
most precise delineations of the dialectical structure in which the poetics of
vitalism, and the very discourse of liberalism itself, emerged. As a committed
Royalist, Harvey reveals more readily than any of our writers (with the possible exception of Marvell) the bad conscience that broods over the new
representation of the liberal state. Nowhere do we see Harvey struggling so
powerfully with the ideological implications of his own radical vitalism as in
the treatise on reproduction, published in 1651, two years after his dethroning of the heart.

One of the most important theoretical foundations for Harvey's 1628 theory of the circulation had been Aristotle's argument for the priority of the
heart. The authoritative Galen had, after Aristotle, nominated the liver as
the prince of the body, a conferral of authority that few early modern anat-
omists had attempted to overturn. But Harvey, in the 1628 *De motu*, reasserted the Aristotelian thesis, which bestowed monarchy on the heart on the
basis of that organ's "sensitivity," its possession of "life, motion, and sense"
that qualified it for its role as centralizing agent (p. 100).[63] We have been
justified, I believe, in attributing to factors of discursive and ideological climate the revolution in Harvey's theoretical assignment of the relation of
heart to blood. But one Harvey scholar has argued that the organizational

63. On the role of Aristotelian "sensitivity" in Harvey's thesis of 1628, see Whitteridge,
William Harvey, p. 221.

change of heart Harvey underwent in 1649 had its roots not in ideology but in an empirical conclusion the doctor drew from an actual medical experiment.[64] It is unlikely, to be sure, that so discursive an event as the dethroning of the heart could have an unmediated origin in empirical observation. But I suggest that we can profit nonetheless from a reading of Harvey's literary account of the experiment believed to have occasioned his reorganization of the bodily polity: the moment Harvey "realized" that the heart was "insensible," incapable of sensation, and that the soul-filled blood, and not the heart, was "the fountain and author of *Sense, Motion,* and *Life* of the whole."[65] A look at Harvey's account of this experiment—actually two experiments—introduces us to a crucial fact of the literary representation of vitalist agency with which all the following chapters are concerned. The new formulation of independent vitalist agency is founded at its inception on a rhetoric of negation: in Harvey, in Marvell, in Milton, and in Cavendish, the literary expression of liberal agency, at this early moment in liberalism's history, invariably emerges as a dialectical counter to the culturally dominant rhetoric of authoritarian power.

The account of the two experiments Harvey performed, presumably in 1641 but whose implications for the politics of bodily government he does not formulate in print until 1651, exposes, I believe, the extraordinary set of ideological pressures burdening the new organizational philosophy. Two years after publishing the *Circulatio sanguinis,* in a chapter of *De generatione* titled "Of the Blood, as it is the principal part," Harvey declares that he "will not conceale this Admirable *Experiment* (by which it shall appear that the most principal member of all, namely, the very *Heart* it self, may seem to be insensible)" (p. 285). If we had assumed that Harvey's usurpation of "the most principal member" could be dated relatively close to the time of the execution of Charles, the doctor here warns us to situate his theoretical revolution at a moment eight years before the regicide, at the very beginning of the Civil War. Harvey's remarkable narrative is divided into two sections, each of which describes an examination of the young man Hugh, who was soon to succeed to the title of the Viscount Montgomery. The description of the first examination merits, I believe, this lengthy quotation:

> A *Noble* young Gentleman, *Son* and *Heire* to the honourable the *Vice-Count* of *Mountgomery* in *Ireland,* when he was a *childe,* had a strange mishapp by an unexpected *fall,* causing a *Fracture* in the *Ribs* on the *left side.* . . . This

64. Whitteridge, *William Harvey,* pp. 228–29.

65. Harvey, *Anatomical Exercitations,* p. 273. All subsequent quotations from *De generatione* are taken from this earliest English edition, cited in note 21, and noted by page number in the text.

person of Honour, about the eighteenth, or nineteenth year of his *Age*, having been a *Traveller* in *Italy* and *France*, arrived at last at *London*: having all this Time a very wide gap open in his *Breast*, so that you might see and touch his *Lungs* (as it was believed). Which, when it came to the late King *Charles* his ear, being related as a *miracle*, He presently sent me to the Young Gentleman, to inform *Him*, how the matter stood. Well, what happened? When I came neer him, and saw him a sprightly Youth, with a good *complexion*, and *habit* of *body*, I supposed, some body or other had framed an untruth. But having saluted him, as the manner is, and declared unto him the Cause of my *Visit*, by the *Kings Command*, he discovered all to me, and opened the void part of his *left side*, taking off that small *plate*, which he wore to defend it against any blow or outward injury. Where I presently beheld a vast *hole* in his *breast*, into which I could easily put my three Fore-fingers, and my Thumb; and at the first entrance I perceived a certain *fleshy* part sticking out, which was driven in and out by a reciprocal *motion*, whereupon I gently handled it in my hand. Being now amazed at the novelty of the thing, I search[ed] it again and again, and having diligently enough enquired into all, it was evident, that that old and vast *Ulcer* (for want of the help of a skilfull *Physitian*) was miraculously healed, and skinned over with a membrane on the *Inside*, and guarded with *flesh* all about the brimmes or margent of it. But that *fleshy substance* (which at the first sight I conceived to be *proud flesh*, and every body else took to be a *lobe* of the *Lungs*) by its *pulse*, and the differences or *rythme* thereof, or the time which it kept, (and laying one hand upon his *wrest*, and other upon his *heart*) and also by comparing and considering his *Respirations*, I concluded it to be not part of the *Lungs*, but the *Cone* or *Substance* of the *Heart*; which an excrescent fungous *Substance* (as is usual in foul *Ulcers*) had fenced outwardly like a Sconce. The Young Gentlemans *Man* did by dayly warm injections deliver that fleshy accretion from the filth & pollutions which grew about it, and so clapt on the Plate: which was no sooner done, but his *Master* was well, and ready for any *journey* or *exercise*, living a pleasant and secure *life*. (pp. 285–87)

Although Harvey introduces "this wonderful experiment" as one that will reveal that the heart is "insensible," it is not in the account of this initial examination that he makes this discovery known. There is no point in the narration quoted above, not even as Harvey describes the search with three fingers and thumb of poor Montgomery's wound, at which the doctor discloses a suspicion that the boy's heart is incapable of sensation. Though Harvey was throughout his career as anatomist eager to cite as an authority his own close inspection of a physiological specimen, on which, as here, he searched everything "again and again," he has not authorized himself in this instance to sanction this most sensitive of medical discoveries. Harvey's account is charged with a rhetorical deference not to the empirically based scientific method—the hallmark of modern science which the doctor is so often credited with introducing—but to an older institution of hierarchical order denoted here as "manner." We see his deference to "manner," or custom, as he bows, with his courteous salute, to the social superiority of the noble Montgomery and as he turns, as we will see, to the personal authority of the king himself.[66]

The anticipated revelation of Harvey's discovery of the heart's insensitivity is carefully withheld until the narrative of a second examination. In this momentous re-examination of Montgomery, the source of experimental wisdom is not Dr. Harvey but King Charles. Harvey has already acquainted us, in the account cited above, with his acquiescence to monarchic authority: it was "the late King *Charles*" who learned of Montgomery's condition and sent Harvey "to the Young Gentleman, to inform *Him*, how the matter stood"; it was Harvey's assurance that he had come to Montgomery "by the *Kings Command*" that persuaded the modest youth to lay bare the "void part of his *left side*." But it is not until the description of the second experiment that the full implications of this deference make themselves known. Harvey, it would appear, did not trust himself to report to the king his experimental findings; instead of taking back "an *Account* of the *Business*," as, we can assume, he typically did, Harvey arranged for Charles's own examination of Montgomery's body:

> I brought the Young Gentleman himself to our late *King*, that he might see, and handle this strange and singular Accident with his own *Senses*; namely the *Heart* and its *Ventricles* in their own *pulsation*, in a young, and sprightly Gentleman, without offense to him: Whereupon the *King* himself consented with me, That the *Heart* is deprived of the *Sense* of *Feeling*. For the Party perceived not that we touched him at all, but meerly by seeing us, or by the *sensation* of the outward *skin*. (p. 287)

Only once Harvey narrates this event of the laying on of the king's hands is the central insight of the experiment disclosed: "Whereupon the *King* himself consented with me, That the *Heart* is deprived of the *Sense* of *Feeling*." The king's sight and, perhaps more important, the king's touch are the essential components of the authorization of Harvey's discovery of the heart's

---

66. The conflicted strategies by which Harvey represents the authorization of scientific fact in this treatise look ahead to the Restoration debates about empirical and speculative knowledge between Robert Boyle and Thomas Hobbes, the subject of Shapin and Schaffer's *Leviathan and the Air-Pump*. For a different reading of the period's rhetoric of authoritative knowledge, see Kevin Dunn, *Pretexts of Authority: The Rhetoric of Authorship in the Renaissance Preface* (Stanford: Stanford University Press, 1994), pp. 125–45.

insensitivity: Montgomery's heart is exposed before the king so that His Majesty might "see, and handle this strange and singular Accident with his own *Senses.*" Though Harvey maintains, on the level of narrative, the relation of Charles and Montgomery as that of king and subject, his emphasis on Charles's *"Senses"* draws the king into an intimate rhetorical identity with the exposed heart whose own capacity for sense is precisely the phenomenon in question.

In dedicating, in 1628, his *De motu cordis* to the king, Harvey had asked his monarch to gaze on the human heart: "contemplate the Principle of Mans Body, and the Image of your Kingly power" (p. viii). More than twenty years later, in *De generatione*, Harvey relates having made the same request in person, reaffirming the symbolic identity of heart and king he himself had helped cement. Given that much of that celebrated text of 1628 had established, through both explicit analogy and embedded logic, an inextricable link between heart and king, the exposure of the heart's insensitivity—the discovery that it could not possibly exercise the kind of control over the body that Harvey had once imagined—was decidedly overdetermined. And it is a measure of the ideological burden here that this climactic moment of contact between the prince of the body and the prince of the body politic is framed as a paradox: the king tests this heart with the *"Sense of Feeling"* in his own monarchic hands only to learn that the monarchic heart, his analogical double, is in possession of no such sense. Given the potential organizational implications of the revelation of the heart's incapacity for rule, we should not be surprised that Harvey would permit his theoretical conclusion to be voiced and verified only by the king himself. In *Paradise Lost*, it is only Milton's God who is authorized to reveal the extent of the devolution of his own sovereign power onto his creatures: "no Decree of mine" could "touch with lightest moment of impulse / His free Will" (10.414–46). Similarly, in Harvey's text, it is only Charles who can authorize a medical revelation as symbolically disempowering as that of the heart's inability to feel: it is by monarchic decree that we learn that no decree of the monarchic heart, now proven to be insensitive, can touch with lightest moment of impulse the free and vital fluids of the human body.

Outside the frame of Harvey's narrative, we have been told, the political tensions in 1641, just a year before the critical battle at Edgehill, made themselves manifest even in this medical encounter between a young aristocrat and his sovereign. Charles was reported to have said to the young Montgomery, on pulling his fingers from the cavity in his chest, "Sir, I wish I could perceive the thought of some of my nobilities' hearts as I have seen your heart."[67] Harvey, however, has struggled to keep his own account of

67. Quoted in Whitteridge, *William Harvey*, p. 235.

the meeting free from such open acknowledgments of the king's political vulnerability. Harvey simply informs us that in the years following the king's examination, Montgomery, continuing his daily practice of cleaning and covering his wound, enjoyed "a pleasant and secure *life*." This sanguine medical update, however, exposes an attempt to erase the violence of the political revolution whose symbolic foundations are laid in this very narrative. In his protestation of Montgomery's subsequent well-being, the doctor carefully omits from this 1651 account the fact that in 1649 Montgomery went on to become the commander-in-chief of the Royalist army in Ulster, where he surrendered to Cromwell and was eventually exiled to the Continent.[68] By the time Harvey publishes his narrative account of this experiment, the loyal nobleman whose heart the king touched had already succumbed to the forces that had also destroyed the king himself.

This experiment demonstrating the heart's lack of sensation reveals the incapacity of the body's "most principal member" to function as a governing center or sentient locus of reason and organization that can control as if by will the principality of the body. Given the intellectual imperative of a totalizing organizational system, this revelation in human anatomy could necessitate a corresponding alteration in the understanding of that system Harrington called "political Anatomy." Through the associative logic endemic to the period's philosophical speculation, what the king witnesses here is nothing other than the collapse of monarchy. If only by the knowledge of the later event of regicide that even the earliest readers would bring to this passage, this carefully scripted experiment, with its inset scene of extraordinary physical intimacy between king and loyal subject, rises, I believe, to the literary status of elegy. In drawing out, with considerable poignancy, his own discovery of the symbolic death of monarchy, Harvey imagines that death to be sanctioned by Charles himself. In the "memorable Scene" of regicide invoked by Andrew Marvell in the "Horatian Ode," Charles, with consummate dignity, is said to have "bow'd his comely Head / Down, as upon a Bed" (63–64). We have in Harvey's story of Montgomery and Charles, already denoted here as the "late king," no less gracious an act of monarchic abdication. Like Marvell, Harvey struggles, with an unavoidably ironized pacifist rhetoric, to reconfigure regicide as resignation.

One of the most potent emblems of medieval kingship that survived into the seventeenth century was the doctrine of the thaumaturgical effect of the "royal touch." This principle of the monarch's magic ability to heal the scrof-

---

68. See Whitteridge, *William Harvey*, p. 235. George Ent, in his "Epistle Dedicatory," relays that Harvey confessed that he had labored over *De generatione* at a time "when the Commonwealth [was] surrounded with intestine troubles." The "vacation from publique Cares" that was Harvey's experimental research and composition "provided a Sovereign Remedy" to his mind (*Anatomical Exercitations*, sig. A3ᵛ).

ulous had become in the mid-seventeenth century an ideological touchstone for both sides in the Civil War. Belief in the royal touch was by the 1640s "one of the dogmas of the royalist faith rejected by the supporters of the Long Parliament."[69] When, in 1642, Charles was compelled by war to move his court to Oxford, royalist propagandists circulated pamphlets entreating him to return to London, where the sick could be cured by those "supernaturall meanes of cure which is inherent in your sacred Majesty." Exploiting the political value of the royal touch, Charles's loyal petitioners asked the king to return to London not only to heal the sick but to heal "the State, which hath languished of a tedious sicknesse since your Highnesse departure."[70] In his own description of the king's touch, Harvey makes clear that whatever curative miracle was available to Montgomery had occurred long before his physical contact with the king: "it was evident, that that old and vast *Ulcer* (for want of the help of a skilfull *Physitian*) was miraculously healed." Incapable of healing the young nobleman's wound, Charles was also unable to heal the wound of State, the larger cultural wound that at this pivotal moment in English history had rendered it impossible, even for a loyal subject like Harvey, to imagine the relations among people, things, and parts of the body as naturally and inherently monarchic. Charles, to be sure, is credited with the singular, perhaps even sacred, ability to validate Harvey's experimental findings. But what the king has authorized is not only his own symbolic demise but the destruction of an entire tradition of organizational philosophy that required the presence of a single sentient figure at the organizational center of all bodies and perhaps, too, of all bodies politic. It is as if, by "consenting" to Harvey's dethronement of the heart, Charles has acquiesced to an organizational philosophy founded not merely on the liberal principle of consent, but on the particular, quite radical, form of consent a constitutional monarch owes to the dominant will of his subjects.

It is impossible to determine the degree of deliberate literariness behind Harvey's account of this experiment with the young Montgomery, though Harvey's friend, George Ent, describes this treatise as a "work framed and polished with very great pains" ("Epistle Dedicatory," n.p.). But the text nonetheless makes manifest the antithetical outlines of the discursive struggle in which all the authors treated in this book are engaged. Harvey labors, in his late writings on the heart, to represent a vitalism that counters both the rule of force in a mechanist such as Descartes and his own mystically

---

69. See Marc Bloch, *The Royal Touch: Sacred Monarchy and Scrofula in England and France,* trans. J. E. Anderson (London: Routledge, 1973), p. 209.

70. *To the Kings most Excellent Majesty The Humble Petition Of divers hundreds of the Kings poore Subjects* (London, 1643), p. 8; quoted in Bloch, *Royal Touch,* p. 209.

founded authoritarianism of 1628. He articulates a theory of independent agency that shifts the power of action and organization from the center to the periphery. But at this fascinating juncture before the entrenchment of a full-fledged liberal discourse, this expression of liberal independence emerges dialectically, in the context of the doctor's simultaneous acceptance and rejection of the sanctioned paradigms of authority and control.

I want now to turn, in conclusion, to a different question raised by this crucial experimental moment in Harvey's *On Generation*. We have not yet determined why Harvey relates this experiment, so important for his thesis on the blood's agency in the circulation, in this later treatise, *De generatione*, and not in the 1649 *Circulatio*, which would have been its logical home. We cannot of course know absolutely whether the details of the experiment are lifted directly from Harvey's notes of 1641 or whether they are the product of a more considered reconstruction nearer the moment of their postregicidal publication. But we can say with confidence that the literary representation of this case is of such a high level of suggestiveness that it invites a speculative analysis that looks beyond a simple disposition of facts. Harvey's mention of the young Montgomery's "unexpected *fall*, causing a *Fracture* in the *Ribs* on the *left side*," inscribes this text within the almost inescapable seventeenth-century practice of considering all crises, whether intellectual, moral, or political, in the ideologically loaded language of the Fall. We have, in Scripture, a figural precedent for this narrative of an extraordinary event of a young man's fractured rib and the wide wound opening up his left side. I refer of course to the story, in Genesis 2, of the generation, from the left side of Adam, of Eve.

Whereas in the Yahwist account of creation, the first medical procedure and its wonderful result precede the Fall, in Harvey's exploratory restaging of the event, it is the "unexpected *fall*" that causes the "*Fracture* in the *Ribs* on the *left side*." Harvey's reversal of the sequence of birthing and Fall, though a necessary accommodation to the facts of Montgomery's case, also suggests the crucial fact of vitalist representation: the discourse of liberal agency can be generated in this period only in the representational context of the fall or destruction of a prior image of authoritarian agency. Montgomery's fall results in a "very wide gap" that opens up his left side, and it is from deep within this Adamic womb that Harvey fashions a being as "strange and singular" almost as Eve. The birth represented before us, I propose, is the difficult birth of liberalism and, more specifically, of that iconic unit of the liberal polity known as the "individual." At the very moment Harvey witnesses the fall of the authoritarian logic of the body's centralized organization, he assists in the larger cultural delivery of a new category of person, the individual, whose existence is logically dependent on that fall.

Each of the following chapters examines the productive role in mid-seventeenth-century literary discourse of rhetorical contradiction and thematic opposition, and especially the cultural uses to which the systematic literary juxtaposition of incompatible models of agency and organization could be put. The change of heart Harvey underwent from 1628 to 1649 reveals more, surely, than lapses in logic besetting a seventy-year-old physician. Harvey's contradictory representations of cause and structure chart precisely the process whereby the seventeenth-century discovery of the individual was made. The blood, the autonomous "individual" of Harvey's 1649 *Circulatio*, can only be empowered with self-motion, and can only be imagined to govern itself within a nonauthoritarian, liberal organization, once it has seized the capacity for motion and governance formerly attributed exclusively to the heart. The articulation of individualism, at the historical juncture of the Vitalist Moment, is unthinkable without the established paradigm of absolutism, because it is the dispersal of absolute power to all the elements in a system that is the necessary conceptual origin of liberal individualism.

When Milton's Adam gives birth to Eve, "Wide was the wound" we are told, "But suddenly with flesh fill'd up and healed" (8.468–69). When Harvey and the king force the delivery of a new philosophy from the exposed heart of an English aristocrat, they find that his wound had been "miraculously healed" and that they could search it "without offense to him." However painless Milton and Harvey make these individual births appear, a theoretical principle of individualism does not emerge in this period without a significant struggle, a struggle best marked, I think, by this very word's troubled etymology. It cannot be denied that other philosophies of material organization—the mechanistic hypothesis in particular—had a claim on the birth of the "individual." The mechanical theory of the isolable atom, derived ultimately from the corpuscular philosophies of Democritus and Epicurus, played an important role in the contemporary cultural meditations on the constitution of society. The infinitesimal unit of the mechanist's physical universe, the "atom," provided, I believe, one of the figural bases for the central element of the social universe, the "individual," a category of increasing interest to intellectual disciplines as diverse as Arminian theology and the nascent philosophy of political consensus. The word *individual* was just beginning in the later seventeenth century to acquire its status as a noun, and its inextricable ties to a specifically mechanistic science of matter are perhaps made all the clearer by the traditional use of the Latin *individuum* for "atom" (from the Greek *atomos*, "undividable"). The lifeless and inert atoms of matter provided for many in this period, most notably Hobbes, tempting analogues for individual human beings. Like change in the social world, change in the natural world could be viewed in the later seventeenth century as the consequence of the violent, almost purposeless collisions of atomic particles, all of which were necessarily overseen, at some ultimate point, by an arbitrary sovereign.

Such, however, was only one of the ways in which the "power of matter" exerted itself on the formation of the conflicted principle of the individual and the liberalism it subtended. There was, I believe, another vision of the "power of matter" that made a competing etymological claim on this word that would later establish itself as the sanctioned signifier for a member of the polity. That alternative materialism was, of course, vitalism. Harvey, Milton, Cavendish, and, though in a less theoretical fashion, Marvell all reveal their commitments to this alternative vitalist genealogy of the individual by the strength with which they propound the monistic infusion of body with spirit. Milton's assertion of monism is the most forceful: "Man is a living being, intrinsically and properly one and individual [*individuum*], not compound or separable, not, according to the common opinion, made up and framed of two distinct and different natures, as of soul and body, but that the whole man is soul, and the soul man, that is to say, a body, or substance individual [*individuam*], animated, sensitive, and rational."[71] "Individual" used adjectivally in this nonmechanical monistic context points not to the inert indivisibility of the lifeless atom, cast about by force and a governing providence, but to a philosophy of a material body from which life, sense, and reason can never be divided. The new English substantive "individual" was perhaps initially a metaphysician's simple shorthand for "individual body," the monistic union to which Ralph Cudworth gestures when discussing the "Compassion which the Soul hath with that Individual Body with which it is Vitally United."[72] It was the specifically vitalist doctrine of the indivisibility of body and soul, I believe, that informed what may have been the first political invocation of the "Individuall," in the claim asserted by one of the period's earliest monistic radicals, the mortalist Leveller Richard Overton: "every Individuall in nature is given an individuall property by nature,

71. *The Works of John Milton*, 18 vols., ed. Frank Patterson et al. (New York: Columbia University Press, 1931–40), 15:41. Further quotations from Milton's prose are drawn from this edition and are cited by volume and page number in the text.

72. Cudworth, in his posthumously published *Treatise Concerning Eternal and Immutable Reality* (London, 1731), pp. 80–81. For a reading of the origin of "individual" that focuses on the word's ties to doctrines of marriage and of the Christian Trinity, see Peter Stallybrass, "Shakespeare, the Individual, and the Text," in *Cultural Studies*, ed. Lawrence Grossberg, Cary Nelson, and Paula A. Treichler (New York: Routledge, 1992), pp. 593-610. See also the brief but important discussion of "individual" in Raymond Williams, *Keywords: A Vocabulary of Culture and Society*, revised ed. (New York: Oxford University Press, 1983), pp. 161–65.

not to be invaded or usurped by any."[73] For the vitalist radical of any political persuasion, the flesh is heir to its own source of power and agency, bequeathed it as "property by nature, not to be . . . usurped by any." In the tentative, dialectical expressions of the liberal individualism we examine in the following chapters of this book, we see the peculiarly corporeal origins of that momentous ideology of personal agency and decentralized organization. We see the faith infused at the Vitalist Moment throughout the disciplines, and, curiously, throughout the political spectrum, in the self-governing integrity of the political body and the body politic.

73. Richard Overton, *An Arrow Against All Tyrants and Tyrany, shot from the Prison of Newgate into the Prerogative Bowels of the Arbitrary House of Lords and all other Usurpers and Tyrants Whatsoever* (London, 1646), p. 3. Overton had two years earlier introduced monistic mortalism to England with his *Mans Mortalitie* (1644).

2 Marvell, Winstanley, and the Natural History of the Green Age

> I say that nothing is made without seed: all things are made by vertue of seed: and let the sons of Art know, that seed is in vain sought for in trees that are cut off, or cut down, because it is found in them only that are green.
>
> —JOHN FRENCH, *A new light of Alchymie* (1650)

With the exception of the poems of Milton, few lyric poems of the seventeenth century have elicited as many critical attempts to discern their ideological significance as the Mower poems and *Upon Appleton House* of Andrew Marvell. In the wake of the famous debate between Cleanth Brooks and Douglas Bush on the question of the historical significance of Marvell's "Horatian Ode," there has prevailed for some years now the critical assumption that Marvell's pastoral poems—"Damon the Mower," "The Mower against Gardens," "The Mower to the Glo-worms," "The Mower's Song," as well as *Upon Appleton House*—cannot fully be understood until they are pressed to render up their historical significance.[1] For the Elizabethan critic George Puttenham, a pastoral poem's historical meaning could be located in the poet's desire to "insinuate and glaunce at greater matters."[2] And those

1. Cleanth Brooks, "Marvell's 'Horatian Ode'" (1946), and Douglas Bush, "Marvell's 'Horatian Ode'" (1952), are both reprinted in Michael Wilding, ed., *Marvell: Modern Judgements* (London: Macmillan, 1969), pp. 93–124.

2. George Puttenham, *The Arte of English Poesie*, ed. Gladys D. Willcock and Alice Walker (Cambridge: Cambridge University Press, 1936), p. 38. For political readings of Marvell's pastorals, see, for example, Don Cameron Allen, *Image and Meaning: Metaphoric Traditions in Renaissance Poetry*, 2d ed. (Baltimore: Johns Hopkins University Press, 1968), pp. 187–225; and more recently, Michael Wilding, *Dragons Teeth: Literature in the English Revolution* (Oxford: Clarendon, 1987), pp. 138–72. The sociopolitical readings of Marvell's images of agrarian labor, an interpretive tradition I discuss later in this chapter, include Raymond Williams, *The Country and the City* (New York: Oxford University Press, 1973), pp. 26–34; Anthony Low, *The Georgic Revolution* (Princeton: Princeton University Press, 1985); Annabel Patterson, "Pastoral versus Georgic: The Politics of Virgilian Quotation," in *Renaissance Genres: Essays on Theory, History, and Interpretation*, ed. Barbara Kiefer Lewalski (Cambridge: Harvard University Press, 1986), pp. 241–67; and Rosemary Kegl, "'Joyning my Labour to my Pain': The Politics of Labor in Marvell's Mower Poems," in *Soliciting Interpretation: Literary Theory and Seventeenth-Century*

critics of Marvell who have attempted to supply a historical understanding of the pastorals have typically proceeded on just this assumption of the possibility of topical analysis. There can be little question that many of Marvell's poems glance, however coyly, at an identifiable world of seventeenth-century persons and events. But in analyzing the relation between Marvell's pastoral poetry and what may seem to be the "greater matter" of contemporary political history, I hope to do more in this chapter than merely unveil Marvell's allegorizations of midcentury politics or revolutionary theology. It is not precisely "history" as the aggregate of particular past events, whether recent or biblical, or even as the array of current political or socioeconomic affairs, that I see as the ultimate contextual frame of Marvell's pastoral poems. It is, rather, an abstract but no less pointed discourse of political agency and natural philosophical process, a discourse specific to the Vitalist Moment, that supplies the resonance of historical urgency to Marvell's pastoral poems. The Marvellian poetic is a discursive structure at once immersed in and amused by the vitalist refiguration of agency undertaken in the very years of the poet's patronage by Thomas, Lord Fairfax. Marvell's representation of pastoral action bears a particularly close relation to the theologically and alchemically inflected representations of agrarian labor by his contemporary, the vitalist communist visionary Gerrard Winstanley. In redrawing the lines of analysis by which the historical referentiality of a midcentury lyric poem can be identified, we can arrive not only at a new understanding of Marvell's pastoral poems. We can arrive as well, by the end of the chapter, at a new, historically informed understanding of the justly celebrated literary phenomenon of Marvellian irony.

The Diggers and the Evasion of Historical Agency

It is Marvell's patron of the early 1650s, the former Lord General Thomas Fairfax, who supplies the most obvious historical justification for the discursive alliance I am proposing between the poet and his vitalist contemporary, Gerrard Winstanley. Two months after the execution of Charles, on April 1, 1649, Winstanley and a group of like-minded radicals began to cut down trees and cultivate the commons on St. George's Hill in Surrey.[3] Articulating a critique of the very principle of wage labor, and of the laborer's rent on a

---

English Poetry, ed. Elizabeth D. Harvey and Katharine Eisaman Maus (Chicago: University of Chicago Press, 1990), pp. 89–118.

3. Winthrop S. Hudson, "Economic and Social Thought of Gerrard Winstanley: Was He a Seventeenth-Century Marxist?" Journal of Modern History 18 (1946): 1–2.

"commons" that had come under the control of wealthy landowners, these "Diggers," as Winstanley and his followers were called, were laboring to establish the first socialist utopia with aspirations of national political reform.[4] The Diggers, in their attempt to achieve this reform, encountered considerable opposition, not least from the Lord General of the Parliamentary army. Initially, the Diggers stood firm. After confronting a disgruntled Lord Fairfax personally that year, once on April 20 at Whitehall and once on May 26 on St. George's Hill, Winstanley reiterated his belief that "the common People ought to dig, plow, plant and dwell upon the Commons, without . . . paying Rent to any."[5] But throughout 1650, the Council of State directed General Fairfax to send troops in aid of local magistrates who found their villages unsettled by the bands of Diggers who were establishing agrarian communes on the local commons. Lord Fairfax, who was surely appalled by the ideological threat posed by Winstanley's communist idealism, was nonetheless reluctant to disperse or destroy the Digger community with force. An official deployment of force, as it happened, was not required. Outbursts of protest and even violence by ad hoc vigilantes in nearby communities had successfully dissipated all the Digger colonies by the next year, 1650, the same year the thirty-six-year-old Fairfax would retire from his duties as Lord General of the Parliamentary forces. The demise of this short-lived group of political activists coincided not only with the demise of Fairfax's military career but with the commencement of Marvell's residence at Nunappleton as tutor to Fairfax's daughter, Mary.

The careers and personal concerns of both Fairfax and Winstanley were marked in 1650, at least on the surface, by the quick slide from political action to embarrassed retirement. It is perhaps the surprising complexity of the notions of action and retirement as they are applied to the career of either Fairfax or Winstanley that drew Marvell to consider these two very different political figures in the same literary work. After disappearing from St. George's Hill in 1650, the Diggers resurfaced, on the plane of literary representation, in Marvell's paean to Lord Fairfax and his estate, *Upon Appleton House*. Christopher Hill, Michael Wilding, and others have identified the "Levellers" invoked in line 450 of that poem as the Diggers.[6] Winstanley had styled his group the "True Levellers," at once associating and dissociating his movement with the democratic Levellers of the army, whose goal of parliamentary representation for property owners no doubt provided an

4. Sabine, ed., *Works of Winstanley*, p. 5.

5. Ibid., pp. 16–18. See Thomas N. Corns, *Uncloistered Virtue: English Political Literature, 1640–1660* (Oxford: Clarendon, 1992), p. 165.

6. Christopher Hill, *The Religion of Gerrard Winstanley* (*Past and Present*, supplement no. 5, 1978), p. 52; Wilding, *Dragons Teeth*, p. 153.

enabling revolutionary logic for Winstanley's far more radical social vision. The contempt in which Lord Fairfax must have held these "True Levellers," who did not even believe, like the army's Levellers, in the principle of private property, can be heard with little effort in one of the most politically mean-spirited stanzas in Marvell's poem. After the speaker's apocalyptic vision of the harvest in *Upon Appleton House*, with its strong emblematic suggestion of Civil War battlegrounds, we are invited to see the spare vegetation of the estate fields as a canvas ready for a fresh representation: "A levell'd space, as smooth and plain, / As Cloths for *Lilly* strecht to stain." This leveled space, the speaker tells us, is that

> naked equal Flat,
> Which *Levellers* take Pattern at,
> The Villagers in common chase
> Their Cattle, which it closer rase;
> And what below the Sith increast
> Is pincht yet nearer by the Beast.
> Such, in the painted World, appear'd
> *Davenant* with th' Universal Heard.
>   (441-56)[7]

It is not difficult to discern in this stanza the looming political threat Marvell attributes to these villagers, whose grazing of cattle on Fairfax's estate assumes the shape of the threat to property posed by the Diggers. Stripping the field of vegetation, they resemble the True Levellers, who in their eagerness to recover the "World when first created" sought to reduce the world to a "Table rase and pure." Marvell alludes to the flip ekphrastic hexameron in Davenant's *Gondibert* to suggest that the local villagers are as grotesquely ungovernable as Davenant's Edenic cattle before the creation of man: "Wondring with levell'd Eyes, and lifted Eares, / Then play, whilst yet their Tyrant is unmade."[8] The hint of animosity in Marvell's stanza toward the possible radical design of herdlike commoners is understandable. *Upon Appleton House* is, to a large extent, Marvell's celebration of his patron's private property; and private ownership was a legal category to which the Diggers referred frequently as an act of "high treason against the King of Righteousness."[9]

7. All quotations from Marvell's poetry are taken from *The Poems and Letters of Andrew Marvell*, 2 vols., ed. H. M. Margoliouth, rev. Pierre Legouis (Oxford: Clarendon, 1971), vol. 1, and are cited by line number in the text.

8. *Sir William Davenant's Gondibert*, ed. David F. Gladish (Oxford: Clarendon, 1971), p. 166.

9. Gerrard Winstanley, *The New Law of Righteousness*, in *Works of Winstanley*, p. 201.

Marvell would most certainly have shrunk from the social and political implications of Winstanley's revolutionary agenda. Despite the celebrated ambivalence with which Marvell caressed the pressing matters of Charles's execution or the rise of Cromwell, there is no literary evidence for the poet's occulted sympathy with one of the most radical and uncompromising political movements of the revolutionary period. It is not, however, a hidden commitment to Winstanley's leftist politics that I am seeking to establish. In this reading of Marvell's political poetic, I wish to identify a distinctive strain of the rhetoric of monistic vitalism, the belief in the union of body and motive spirit that finds a theological and political manifestation in the prophetic rhapsodies of the Digger Winstanley. Marvell shares with Winstanley, with a few other Puritan radicals, and with vitalist philosophers such as William Harvey and Jean Baptiste van Helmont a conceptualization of agency that defies the orthodoxies of providential Calvinism and mechanistic Hobbesianism through an engagement of a revolutionary doctrine of animist materialism. The tradition of literary pastoral had always made room for various manifestations of the pathetic fallacy. But Marvell imports into his pastoral poetry a culturally specific form of vitalist materialism that bore, as we saw in Chapter 1, a range of radical political associations in the early years of the Interregnum. If Marvell cannot be seen to share Winstanley's critique of property rights, there is, I think, an unusual and surprising sense in which the "levell'd space" of Marvell's poetic canvas reveals a representation of Winstanley's complex articulation of the process of revolution, the Digger's literary account of how reformation actually occurs.

In his most original and daring treatise, *Fire in the Bush* (1650), Winstanley uses (as, we will see, Marvell does) the pivotal moments of Christian history to structure his forecast of the trajectory of England's long-awaited social and political reformation.[10] The scriptural event of the Fall Winstanley reads as the expropriation of man, by selfish landlords, from the communal property of the garden. The gradual realization of a social utopia is figured throughout his work, then, as a return to the Edenic state of communality. But the Eden for which Winstanley yearns is at the same time more than a political state redrawn along communistic lines; it is an egalitarian state curiously moral and even physiological in its outlines. The social conditions he attributes to the unfallen garden are inseparably attuned to the natural organic environment in which the unfallen human body finds its physical home.

---

10. See the examination of *Fire in the Bush* in T. Wilson Hayes, *Winstanley the Digger: A Literary Analysis of Radical Ideas in the English Revolution* (Cambridge: Harvard University Press, 1979), pp. 174–219.

This radical conflation of creature and creation makes all the more potent his understanding of the Fall from communal property, which is, for him, the individual's "expulsion" from a consciousness of the divinity that dwells within his flesh. Winstanley's revolutionary agenda, then, involves the attempt to regain a lost paradise, to reunite in a startlingly literal and hermetic sense the body of the human animal to the land and vegetation that surrounds it. The anticipated revolution is continually figured as a vegetabilization of humanity, a process that is a coming into what Winstanley calls the "Life of Pure Reason," for "Pure Reason lives like a corne or wheat, under the clods of Earth."[11]

It is the nature of this political redemption, and of the peculiar mode of apocalypse that effects it, that sets Winstanley apart from the apocalyptic Puritans with whom he was contemporary. When in *Fire in the Bush* he imagines the eschatological destruction of the evils of private property and enclosure, he continually recalls God's curse on the serpent in Genesis 3:15, the *protoevangelium* that enjoyed a privileged status among midcentury radicals as a harbinger of revolutionary change: the woman's seed shall bruise the serpent's head. In appropriating this scriptural prophecy for his own political vision, Winstanley transmogrifies the "woman's seed" of Genesis into a seed from which will sprout and flourish the actual Tree of Life, the tree mentioned in Genesis 3:22 as the Edenic plant that possessed the power to give man immortality and for the sake of whose presence Adam and Eve had to be driven from the garden. This Tree of Life, or "Seed of Life," is for Winstanley still alive and capable of returning the earth to its Edenic state through a process surprisingly mundane and accessible: "This Tree of Life I say, is universall Love, which our age calls righteous Conscience, or pure Reason; or the Seed of life that lies under the clods of Earth, which in his time is now rising up to bruise the Serpents head, and to cast that imaginary murderer out of the Creation."[12] We have here a characteristic example of the powerfully literalizing quality of Winstanley's figurative drive, his persistent return of a theological abstraction to the agrarian world of cultivation that forms the basis of the Digger utopia.[13] Winstanley begins this description of the mainspring of the Digger revolution by identifying the redemptive force of the "Tree of Life" with the abstract principle of "Love," a principle that undergoes further amplification when he notes that this is the Love "which our age calls . . . Conscience" and "Reason." As soon, how-

---

11. *Works of Winstanley*, p. 86.

12. Ibid., p. 453.

13. Corns comments on the importance of Winstanley's literalism for his materialist millenarianism in *Uncloistered Virtue*, pp. 156–57.

ever, as Winstanley has sufficiently mystified his sense of the Tree of Life with this series of abstract appositives, he redirects the entire definition, after the semicolon, with an image of extraordinary representational clarity. In identifying this Tree of Life a second time, Winstanley conjures the picture of a seed in the ground shooting forth a sapling tree or a sprig of grass that is capable of fulfilling the promise in Genesis by bruising, literally, a serpent's head, a feat not difficult for the burgeoning sprig, because a serpent's head lies, obviously, so low to the ground. It is this seed of the Tree of Life, writes Winstanley, and not, significantly, an anthropomorphized Christ or even an English soldier, that is "the true and faithfull Leveller."[14]

The murderous serpent of private property will be bruised not by direct human action or aggressive acts of political leveling but by a seed of "patience" and "love": "the spirit of love," writes Winstanley, "would ultimately triumph in all men by virtue of its own power and strength."[15] David Mulder, in his study of the influence on Winstanley of the alchemical writers of the vitalist period, has demonstrated persuasively the tendency in Winstanley to explain the causal efficacy of moral ideals such as "love" and "patience" by an appeal to the principle of the microcosm established first by Paracelsus and developed later by sixteenth- and early-seventeenth-century speculative alchemists.[16] The very structure of the cosmos for Winstanley is founded on the alchemical equilibrium, or temper, of the four elements, fire, water, earth, and air. And this cosmic temper is magically calibrated to the human individual's own fragile elemental equilibrium. The moral and spiritual temper an individual cultivates in himself bears a profoundly literal, material impact on the physical world surrounding him; a state of moral purity or turpitude possesses what Walter Charleton called a "Radiall Activity" in its ability to communicate its corporeal reality to the outside world.[17] Through the mechanism of this deeply osmotic commerce between the human body and the world around it, Adam's sin had its terrible global impact: "when the first man fell, he corrupted the whole creation, fire, water, earth, and

14. *Works of Winstanley*, p. 454.

15. Winstanley, quoted in Mulder, *Alchemy of Revolution*, p. 5.

16. Mulder, *Alchemy of Revolution*, esp. pp. 47–71. Hill, in "Harvey and the Idea of Monarchy," p. 168 n. 42, suggests that Winstanley had read some of the late medical treatises of Harvey and had been influenced by their seeming egalitarianism.

17. "I am bound to believe that in the infinite Magazine of Nature are to be found various Agents, not obliged to the dull condition of an immediate Corporeal Contact; but richly endowed with an Influentiall or Radiall Activity . . . [transmitted] in a semi-immaterial thread of Atomes" (*Ternary of Paradoxes. Translated, Illustrated, and Ampliated by Walter Charleton* [London, 1650], sig. D4ᵛ–E1ʳ; quoted in Nina Rattner Gelbart, "The Intellectual Development of Walter Charleton," *Ambix* 18 [1971]: 155). Henry discusses the thesis of radial activity in "Occult Qualities," p. 341.

aire, and still as the branches of his body went to the earth, the creation was more and more corrupted, by the multiplicity of bodies."[18] Because of the peculiarly physiological interflux between the creature and creation, the world suffered the evil of the Fall as soon as the "branches" of Adam's body were laid in the earth. But the "Radiall Activity" of bodily temper can work as efficiently for good as for ill. And it is by means of this same elemental interaction of human and earthly substance that an individual's cultivation of the bodily virtues of "love" and "patience" can exercise a material, indeed revolutionary, impact on the outside world.

Winstanley brings the alchemical principle of elemental temper into contact with the related alchemical principle of the seed, the mystical agent of germination without which, as John French, a Paracelsian physician from Fairfax's army, had written in the epigraph quoted above, "nothing is made." Through his appeal to the alchemist's "seed," Winstanley signals his commitment to the vitalist inversion of causes and effects, agents and patients: he performs the difficult and fascinating maneuver of making passive a scriptural event that by any other interpretation would require some type of direct human action. The phrase from the verse in Genesis noted above—the "woman's seed"—had always suggested to readers of the Bible a human agent, or, if a divine agent, one, like Christ, with human attributes. In Winstanley's hands, however, the scriptural metaphor of the woman's seed is returned to its literal root and subjected to a back-formation into an impersonal force of natural growth. The seed, the origin of the Tree of Life, is without question for Winstanley a powerful instrument of revolution. But the seed has no claim to a transcendent realm of immaterial agency, embodying the same law of nature that Winstanley claims "doth move both man and beast in their actions, or that causes grass, trees, corn and all plants to grow in their several seasons."[19] It is a vegetable love, implies Winstanley, that can grow vaster than empires, creating of its own accord a universal empire of communal property. Peter Elmer has persuasively argued for the importance of a sectarian medical movement, thriving in the 1640s and 1650s, that he has identified as "eirenicism." The loosely aligned eirenic intellectuals, favoring ecclesiastical toleration and universal peace, were often, Elmer demonstrates, "drawn from the ranks of hermetic and chemical physicians."[20] With his insistence, throughout the Digger tracts of 1649 and

18. Winstanley, Works of Winstanley, p. 115; quoted in Mulder, Alchemy of Revolution, p. 56.

19. Gerrard Winstanley, The Law of Freedom, quoted in Lewis H. Berens, The Digger Movement in the Days of the Commonwealth (London: Simpkin, Marshall, Hamilton, Kent, and Co., 1906), pp. 222–23.

20. Elmer, "Medicine, Religion, and the Puritan Revolution," p. 34.

1650, on a historical force more alchemical, even botanical, than human, Winstanley forges a voice particularly suited to the eirenicism of the Vitalist Moment, the peculiarly passive voice of the political activist.

The implications of Winstanley's strange hermeneutic might be illuminated if placed alongside an alternative contemporary interpretation of the image of the woman's seed from Genesis 3. Reconciling a millenarian theology with an aggressive political agenda, another revolutionary group of the 1650s, the Fifth Monarchists, grounded their stubbornly noneirenic practice in a very different belief that the only way to precipitate the establishment of the Fifth Monarchy of Christ was through the violent overthrow of the state. As Nigel Smith has observed, the Fifth Monarchists could justify their militancy with an interpretation of just that figure of the "seed" so important to Winstanley's politics of passivity. The Fifth Monarchist Christopher Feake, for example, in his pamphlet *A Beam of Light*, "referred to the Army as the Seed, or the bearer of the Seed, so that the promise of regeneration is contained within the actual force of reformation."[21] In identifying the "seed" with the army, Feake merely locates the human referent logically implied in the *protoevangelium*. It is with an unquestionable hermeneutic justice that he returns the image to the implicitly activist agent suggested by the Genesis original. The reasonableness of Feake's interpretation should serve to highlight the distinctive oddity of Winstanley's reliance, for the purposes of his passive revolution, on a text that practically demands an activist reading. Winstanley's erasure of human action from the dynamics of political revolution is manifest in a gesture of interpretive distortion so extravagant that it cannot help but expose the self-contradictory energies charging his work, the conflicts that emerge throughout his writing between natural and human modes of revolutionary agency.

Here we must address the concern with these conflicts that has been at the center of the scholarly controversies surrounding Winstanley and the Digger movement. What, if any, type of political action did Winstanley advocate? What was the *agent* of change for the revolution he was awaiting? The historians of this movement can be usefully divided into two identifiable camps: those who read Winstanley's political aspirations as recognizably practical and tangible, and those who find his political program hopelessly benighted and mystical.[22] It is Winstanley's understanding of the meaning of

21. Nigel Smith, *Perfection Proclaimed: Language and Literature in English Radical Religion, 1640–1660* (Oxford: Clarendon, 1989), p. 242.

22. David W. Petegorsky, *Left-Wing Democracy in the English Civil War: A Study of the Social Philosophy of Gerrard Winstanley* (London: Gollancz, 1940), and J. C. Davis, "Gerrard Winstanley and the Restoration of True Magistracy," *Past and Present* 70 (1976): 78–92, argue that Winstanley advocated a genuine political activism. Acknowledging the presence of Win-

his "digging" that Winthrop Hudson properly identified in 1946 as the most mystifying paradox for historians of the movement.[23] There has been no consensus in the years since Hudson's analysis on the question of the value and weight Winstanley accorded that mysterious action. Winstanley frequently figures the activity of his digging as unmediated political intervention: he encouraged his fellow Diggers in the spring of 1650, for example, to "stand up for your freedom in the Land, by acting with *Plow* and *Spade* upon the *Commons*."[24] This "acting with *Plow* and *Spade*" was often interpreted by contemporaries as a direct political threat, as many seemed convinced that the Diggers were working to employ plow and spade to tear down, or level, the lines of trees, boundaries, and fences that marked the lines of property; the "digging," in this view, was a direct means of redistributing the nation's wealth. But Winstanley had personally assured Fairfax that the Diggers would never deploy "strength, or a forcible obstruction" to fulfill their revolutionary goals. He had written repeatedly that he had no intention of expropriating landowners: the common people would clear and cultivate only the common lands, and the mysterious "Seed" from the "Tree of Life," and not a strong-armed group of Diggers, would oversee the eventual dissolution of freeholds, the return of all England to the paradisal state of communal property. The hoped-for revolution, as its trajectory is represented in *Fire in the Bush* and elsewhere, would be the result not of human effort, not even of divine providence, but of an organic process regulated internally by what Winstanley saw as an abstract principle of immanent divinity. Winstanley's protestations of passivity aside, however, the complex activity of digging must still be accounted for: if the communal paradise was to be regained by the forces of a vegetative providence, what, then, was the function of Winstanley's "acting with *Plow* and *Spade*"? The gesture of "digging" has proven, and will continue to prove, an irresistible object for historical interpretation because of its intractably ironic status. The physical act of clearing trees and digging up the commons, a practice frequently repre-

---

stanley's idealist rhetoric, Petegorsky insists nonetheless that the "essence of his doctrine was his realization that social change had to be initiated neither by the spirit of love nor by the force of reason, but only through the direct action of politically conscious individuals" (p. 213). Like George M. Shulman, in *Radicalism and Reverence: The Political Thought of Gerrard Winstanley* (Berkeley: University of California Press, 1989), Mulder, *Alchemy of Revolution*, p. 11, takes seriously Winstanley's idealist rhetoric: "In the end, what we see developing is a curious picture of a revolutionary movement striving to provide for the most basic needs of the very poorest sort of people through what we would call magic." Mulder, pp. 176–77, discusses the tension in Winstanley between idealist and political modes of social change.

23. Hudson, "Gerrard Winstanley," pp. 5–6.

24. *An Appeal to all Englishmen*, in *Works of Winstanley*, pp. 413–14.

sented by Winstanley as aggressive political action, necessarily contradicts the unswerving passivity of official Digger theology.

The aspect of the digging that most clearly troubled Winstanley's contemporaries was the clearing of trees on the commons. It is surely not insignificant for our understanding of the Digger movement, and our understanding of that movement's effect on the poetry of Marvell, that it was at this moment in English history that the country's standing groves of trees had begun to acquire an unmistakable set of nationalistic associations. The early modern conservationist and belle-lettrist John Evelyn had gone so far as to identify the rich resources of the kingdom's forests as the basis for England's military strength: "our forests are undoubtedly the greatest magazines of the wealth and glory of this nation; and our oaks the truest oracles of its perpetuity and happiness, as being the only support of that navigation which makes us fear'd abroad, and flourish at home."[25] This heightened appreciation of the English tree, however, emerged as a response to a more immediately troubling, ideologically loaded phenomenon. One of the most striking consequences of the political turmoil of the 1640s and 1650s was a shocking increase in the illegal felling of trees on both public and private lands. Indeed, "during the Interregnum," according to Keith Thomas, "no actions aroused more passion than the felling of groves on Royalist estates."[26] The "magazine of timber" in England and Ireland, wrote Evelyn, during the Restoration, had been "destroyed by the Cromwellian rebels."[27] To be sure, as Evelyn himself had noted, many trees had been destroyed for practical reasons and justifiable ends: a good deal of the forest had been sacrificed for the expansion of the nation's naval fleet; and the timber on confiscated Royalist estates provided, in addition, a ready source of fuel for those whose access to forests had been curtailed by a spate of recent enclosures.

But trees were also being leveled in this period in more symbolic acts of political protest. The proponents of the antiprelatical Root and Branch Petition had been prepared to publicize their political cause by destroying the famous procession of elms in St. James's Park.[28] And it was the felling of an entire system of social justice that Winstanley was himself proposing in his

---

25. John Evelyn, *Sylva, or a Discourse of Forest Trees*, 2 vols. (1664; London: Doubleday, 1908), 2:157. Samuel Hartlib forwarded a similar position in his 1658 petition for the improvement of forestry, reprinted in Webster, *Great Instauration*, pp. 546–48.

26. Keith Thomas, *Man and the Natural World: A History of the Modern Sensibility* (New York: Pantheon, 1983), pp. 218–19.

27. *Sylva*, 2:169. Margaret Cavendish gives an extensive accounting of the trees destroyed on her husband's confiscated lands in her *Life of the Thrice Noble . . . William Cavendishe* (London, 1667).

28. Thomas, *Man and the Natural World*, p. 219.

surprisingly militaristic tract of 1650, *A Declaration from the poor oppressed people of England*: "the main thing we aym at, and for which we declare our Resolutions to go forth, and act, is this, To lay hold upon, and as we stand in need, to cut and fell, and make the best advantage we can of the Woods and Trees, that grow upon the Commons, To be a stock for our selves, and our poor Brethren, through the land of *England*."[29] When Winstanley calls upon his poor brethren to "cut and fell . . . the Woods and Trees, that grow upon the Commons," he is doing more than laying hold upon those resources of which he stood in need; he is tapping one of his culture's most weighted emblems for political intervention.[30]

Like his radical contemporaries, Winstanley no doubt took comfort in the fact that this particular form of political action was accompanied by a sanction in Scripture. In the Gospel of Matthew, John the Baptist had issued a threat of arboreal sacrifice to warn the corrupt Pharisees and Sadducees that the Christian dispensation would uproot their long-held privilege and position: "And now also the axe is laid unto the root of the trees: therefore every tree which bringeth not forth good fruit is hewn down, and cast into the fire" (Matt. 3:10). But the presence in the Bible of an analogous political symbology is not, I suggest, sufficient to release Winstanley from the logical morass into which his revolutionary rhetoric places him. The scriptural precedent for the imaging of a revolution as the clearing of trees had not emerged in Matthew, as it does in Winstanley, in the context of a thoroughgoing vision of a vegetative reformation. The logical impasse that opens between the cutting down of trees and the cultivation of their growth—between the calls for radical action and the passive faith in a natural reformational process—manifests on a rhetorical level the logical fissure that seems to doom to incoherence the entirety of Winstanley's communist agenda. The Digger's activist rhetoric is desperately at odds with his pacifist theology, which figures an uninterrupted, organic reformation specifically as the cultivation of a tree, the biblical Tree of Life, as the one and only agent of revolution. However practical or useful the Diggers' actual clearing of the wooded commons, the felling of trees, as a symbolic gesture, carries with it a burden of association that quite simply overwhelms the political and theological vision of a meaningful arboreal growth. We know, of course, that the Diggers were dispersed a year after their formation because of the perception of their threat to the social order. But from a less historical, more literary perspective, we might

29. Gerrard Winstanley, *Selected Writings*, ed. Andrew Hopton (London: Aporia, 1989), pp. 28–29.

30. Milton's *Paradise Regained*, for example, figures Satan's temptation to a military revolution as a similar leveling of a tree-filled landscape: "A multitude with spades and Axes arm'd / To lay hills plain, fell woods, or valleys fill" (3.331–32).

justifiably say that the Diggers were victims, from the inception of the movement, not only of community opposition but of their own bad logic, the inescapable irony generated by the self-defeating interaction of their dominant tropes. As Winstanley may himself have had occasion to read in the work of his contemporary, John French, the alchemist's elusive "seed is in vain sought for in trees that are cut off, or cut down, because it is found in them only that are green."[31]

Marvell and the Philosophy of Trees

We have seen the importance for Winstanley of the revolutionary power enrooted in the image of the Tree of Life and in the seed of the Tree of Life. The plant whose trunk and branches are growing slowly but unstintingly to fill the entire world figures forth Winstanley's ideal of the slow organic growth of reformation. For all their differences on the important questions of revolutionary ends, it is the radical organic vitalism we find in Winstanley—a vitalism whose representation has its peculiar roots in the historically overdetermined figure of the tree—that avails itself to Marvell as he figures in his poetry a satisfactory alternative to his age's predominant agents of historical change. Marvell adapts this image of a tree to represent the vitalist hope in a naturally determined revolution. But he also uses it to reproduce, with his customary lyric concision, the logical and ideological tension fracturing Winstanley's work.

Like Adam in the last two books of *Paradise Lost*, the speaker of Marvell's *Upon Appleton House* witnesses a prophetic pageant of world history. Situated comfortably within the forest on General Fairfax's estate, the narrator claims to read "What *Rome, Greece, Palestine*, ere said" in the "light *Mosaick*" of the leaves and feathers he finds scattered in the woods (581–82). But whereas Adam's prophetic vision is the product of the careful accommodations of a divine agent, Marvell's speaker identifies himself as a proponent of a new, more relaxed mode of historical knowledge: he is the "*easie Philosopher, / Among the Birds and Trees*" (561–62). This sylvan historian simply happens on an instructive historiographical masque, a series of spontaneous tableaux featuring local flora and fauna, playing itself out in the woods of his patron's estate. The speaker remarks of his good fortune, "how Chance's better Wit / Could with a Mask my studies hit!" (585–86). Although few of Marvell's pastoral poems are as explicit as *Upon Appleton House* in their incorporation of a scene of historical interpretation, a persistent interest

31. French, *New light of Alchymie*, p. 24.

in historiographical "study" marks a number of his pastoral works. The interest in the question of historical agency that informs, however improbably, Marvell's particular deployment of the pastoral mode finds its figurative and imaginative basis not only in the traditional conceptual discourses of Rome, Greece, or Palestine but more crucially in the easy philosophy of birds and trees, the specifically vitalist natural philosophical discourse that presented itself in the revolutionary period as a new understanding of agency and organization.

The first scene in the natural masque that the forces of chance bestow on the speaker of Marvell's poem reveals two rows of trees in Fairfax's wood. This accidental tableau displays a historiographical paradigm that captures through antithesis the ideological dialectic we have seen destabilizing Winstanley's revolutionary vision, the same dialectic that comes, we will see, to disjoint the thematic unity of much of Marvell's pastoral verse:

> The double Wood of ancient Stocks
> Link'd in so thick, an Union locks,
> It like two *Pedigrees* appears,
> On one hand *Fairfax*, th' other *Veres*:
> Of whom though many fell in War,
> Yet more to Heaven shooting are:
> And, as they Natures Cradle deckt,
> Will in green Age her Hearse expect.
> (489–96)

Most of the trees the speaker surveys emerge from their origins in "Natures Cradle" in a gradual ascent to heaven, expecting as they grow in their domesticated wood the gentle arrival of Nature's "Hearse." The speaker proceeds in his description of the forest to specify the nature of the trees' heavenly assumption:

> When first the Eye this Forrest sees
> It seems indeed as *Wood* not *Trees*:
> As if their Neighbourhood so old
> To one great Trunk them all did mold.
> There the huge Bulk takes place, as ment
> To thrust up a *Fifth Element*.
> (497–502)

Behind this image of the "one great Trunk" of Fairfax's forest lies the doctrine of the great chain of being, the gradual scale along which all living matter rests in an unbroken continuum of earthly and divine substance. In

Paradise Lost, the angel Raphael employs the loaded analogy of a tree—implicitly, I should think, the Tree of Life—to body forth the monistic progression of substance from dense, earthy matter to rarefied, airy spirit:

> So from the root
> Springs lighter the green stalk, from thence the leaves
> More aery, last the bright consummate flow'r
> Spirits odorous breathes.
>
> (5.479–82)

The forest in Marvell's simile of the "double wood," though it resembles Raphael's account of nature's arboriform continuum of matter, provides a more vividly diachronic image of the chain of being. Marvell's simile shifts this time-honored image of nature's ontological unity from a spatial to a temporal axis, transposing the subtle gradations of matter from their timeless position in an immanent natural hierarchy onto the inexorable motions of a historical process, a process more recognizable as the trees' natural growth. This natural ascent that catches the speaker's eye involves an organically unifying progress from "trees" to "wood" and from wood to ethereal quintessence, an ontological progression, like the one described by Raphael, that traces the decidedly linear path of the Christian schemes of salvation and providential fulfillment. Whereas England's Fifth Monarchists had insisted on a violent hastening of providence as they prepared for that fifth and final monarchy of Christ, the trees in the Appleton wood shoot their way to that final *"Fifth Element"* by gentler means. They shoot their way to heaven not with the guns of militant revolutionaries but with the simple vegetative forces that send the branches and shoots of trees ever higher.

Marvell, like Milton, has been ill-served by the modern critical tendency to ascribe all seventeenth-century expressions of providential change to the voluntarist doctrine of God's predestinary manipulation of the course of human history.[32] With the notable exception of Marvell's "Bermudas," his par-

---

32. Two recent discussions of the theological assumptions behind *Upon Appleton House* find in the poem a reduplication of Calvin's theocentric providentialism. Mistaking Marvell's ironic engagement of contemporary history for an unswerving commitment to eschatological doctrine, Margarita Stocker, in *Apocalyptic Marvell: The Second Coming in Seventeenth-Century Poetry* (Sussex: Harvester, 1986), pp. 46–66, interprets *Upon Appleton House* as a carefully encoded allegory of St. John's Revelation. For Stocker, the "scenes of *Upon Appleton House* record historical events as acts in the eschatological drama," and thus the mowing episode reveals Marvell's faith that "the last Mower, Christ, will harvest mankind in the universal death" (pp. 55, 53). See also Marshall Grossman, "Authoring the Boundary: Allegory, Irony, and the Rebus in *Upon Appleton House*," in *"The Muses Common-Weale": Poetry and Politics in the Seventeenth Century*, ed. Claude J. Summers and Ted-Larry Pebworth (Columbia: University of Missouri Press, 1988), pp. 191–206. Arguing that Marvell subordinates the action of *Upon Appleton House*

ody of Puritan utopian providentialism, it is not Marvell's God that moves
and motivates the physical world of his poems. Marvellian nature, like the
"laden house" on Fairfax's estate (49–52), is an organic body that stirs and
sweats, swells and breathes. Marvell's early poetry is the product of a deeply
heterodox and specifically monistic reformulation of the new conception of
self-motivated change we have seen as a natural philosophy in Harvey and
as a sectarian scheme of redemption in Winstanley. More than any other
body of seventeenth-century lyric poetry, the Marvellian pastoral, with its
curious peaches, its gadding vines, its willing nature, evinces a struggle for
a mode of lyric expression that can reconcile the theology of providential
telos with the increasingly vibrant materialist, often alchemical conception
of change as the product of spiritualized matter in motion.[33]

Despite the anti-Aristotelianism of scientific reformers such as Bacon, the
Aristotelian principle of internally motivated, teleological change lay at the
heart of seventeenth-century Paracelsianism and, as a consequence, of
the natural philosophies of many radical political reformers.[34] Paracelsus himself had written that "alchemy means: to carry to its end something that has
not yet been completed."[35] Following this tradition, Marvell charts the motion of natural objects with a modified Aristotelianism; he views form as the
historical *telos* immanent in matter, the "active force that activates the proper
shape of the thing and actuates the process by which it becomes that thing."[36]
The Creation, for many of the alchemical vitalists, was the result of an elaborate alchemical process.[37] And since the *physis* or teleological process of
alchemical development was immanent in the natural world, even the eschatological image of the end of time could be framed within a materialist

to the seamless providentialism of an Augustine or a Calvin, Grossman, in an otherwise subtle
analysis, locates in the poem Marvell's concern with the "self as subjected to the transcendental
script of Providence" (p. 203).

33. For discussions of the magical and alchemical concerns in Marvell's verse, see R. I. V.
Hodge, *Foreshortened Time: Andrew Marvell and Seventeenth-Century Revolutions* (Cambridge:
Brewer, 1978), pp. 68–95; and Lyndy Abraham, *Marvell and Alchemy* (Hants, England: Scolar,
1990).

34. James R. Jacob argues for the radical ideological thrust of some seventeenth-century
defenses of Aristotle (and the consequent attacks on the implicitly royalist science of mechanism), in *Henry Stubbe, Radical Protestantism, and the Early Enlightenment* (Cambridge: Cambridge University Press, 1983), pp. 78–108. Pocock discusses the role of Aristotelian teleology
in the eschatology of the radical saints of the Civil Wars, in *Machiavellian Moment*, pp. 374–
75.

35. *Paracelsus: Selected Writings*, ed. Jolande Jacobi, trans. Norbert Guterman (Princeton:
Princeton University Press, 1951), p. 141.

36. Richard Kroner, *Speculation in Pre-Christian Philosophy* (Philadelphia: Westminster
Press, 1956), pp. 194–95, quoted in Michael Lieb, *The Sinews of Ulysses: Form and Convention
in Milton's Works* (Pittsburgh: Duquesne University Press, 1988), p. 11.

37. See, for example, French, *New light of Alchymie*, pp. 88–90.

figuration of change. Man's ultimate redemption could be imagined as the result of a natural, temporal process, an operation within time rather than the cataclysmic end to all historical change. The possibility of a naturalized process of providential change allows Marvell, as it does Winstanley, to represent the providential end not as a transcendent revelation of new heavens but as the specifically terrestrial millennium Walter Charleton referred to as the "*Chymistry* of the last day."[38] In his analysis of the "material nature of the envisaged millennium" in Winstanley's writings, Thomas Corns has explained that Winstanley's vision "leads not to the destruction of the earth but the filling of the earth."[39] As we have seen, the more fortunate trees on the Nunappleton grounds can expect the arrival of Nature's hearse "in green Age." And it is perhaps this phrase *green age* that best names the naturalizing vision, of both Marvell and Winstanley, of the promised end of Christian history, a final state of spiritual and vegetative union toward which the whole creation groaningly aspires.

However absurd the attention in this poem to the soteriological destiny of Lord Fairfax's trees, Marvell's employment of this vitalist conceit should signal the general solemnity attending this doubled historical vision. Like the green trees in "Upon the Hill and Grove at *Bill-borow*" (33–40) and "A Dialogue between Soul and Body" (44), the plumps of aged trees in the country-house poem possess an indisputably sacred spirit. But it is a spirit of natural divinity curiously vulnerable to the hostile invasions of human impiety. In order to understand fully the extent to which Marvell's poetry engages on an analytic level the conceptual contradictions thwarting a millenarian vision such as Winstanley's, we must look now at this other, highly unnatural element in Marvell's dialectical tableau, the human intervention implicit in the image of those trees in *Upon Appleton House* that will not find themselves shooting to heaven. Like the members of the Fairfax and Vere families who fought in the English Revolution and in earlier wars, many of the trees in the estate's double wood have been felled in battle. Marvell's stanza on the two pedigrees of trees establishes an antithetical relation between those trees shooting to heaven and those trees destroyed by the soldier's ax. The logical implication of this antithesis is that the trees fallen to hostile hands are, quite simply, dead and irredeemable. If we keep in mind the radical and dissenting nature of the sentiments that motivated the practice of arboreal sacrifice, we can begin to understand the intensity with which Marvell's poem is invested in the fate of trees.

38. Walter Charleton, translator's supplement to van Helmont's *Ternary of Paradoxes* (London, 1650), p. 96; quoted in Mendelsohn, "Alchemy and Politics," p. 43.

39. Corns, *Uncloistered Virtue*, p. 157.

The description of the "double Wood," which "like two *Pedigrees* appears," spatializes in emblematic form the two competing figurations of historical agency and political reformation that mark the writing of Winstanley.
On the one hand, the figure of the heaven-bound arbor, embodying a vegetative force of continual development and growth, suggests the potential
for a gradual, organically driven historical progress toward an ultimate state
of "redemption." The picture of the trees felled purposefully by man, on the
other hand, introduces a considerably different, though perhaps more familiar, notion of historical change: the fact of the destruction of these trees
assumes a prior act of militant human assertion, the active attempt by man
to thwart the seeming progress of an ongoing course of natural development.
This latter, more deliberate practice seems for Marvell not only to threaten
natural process but, even more ominously, to jeopardize a future state of
salvation. Whereas the image of trees growing endlessly heavenward promises the redemptive potential of nature itself, the image of trees irredeemably
destroyed by man—whether by soldier or vandal, Digger or shipwright—
implies an argument for the futility of purposive human attempts at change
and political reform.

Here we touch on what is perhaps the most pervasive thematic paradigm
behind Marvell's poem: the intersection of an image of recent English history
with a Christian discourse of the spiritual history of man. Marvell's vision of
the trees' salvation, I would argue, functions as an element in the poem's
larger inscription of a universal Christian history. Despite Marvell's mockery
of the Diggers' attempt to re-create an Edenic social world by eliminating
private property, the larger movement of *Upon Appleton House* involves a
drive, structurally analogous to the trajectory of *Fire in the Bush*, to redeem
a lost paradisal England that Marvell continually figures as having "fallen"
in war. As Maren-Sofie Røstvig has demonstrated, the narrative of *Upon
Appleton House* is structured to a significant degree by some of the most
foundational elements of Christian historiography—the state of innocence,
the Fall, and that final, compensatory event of redemption. We need not, of
course, conclude with Røstvig that Marvell's use of an essentially Christian
vocabulary removes the poem to the realm of religious allegory or that it
suggests "the existence of an anagogical interpretation leading to a perception of absolute truth."[40] The recurrent images of innocence, Fall, and re-

40. Maren-Sofie Røstvig, " 'Upon Appleton House' and the Universal History of Man," *English Studies* (Netherlands) 42 (1961): 338. A more tempered review of the typological construction of the poem can be found in Barbara Kiefer Lewalski, *Donne's "Anniversaries" and the
Poetry of Praise: The Creation of a Symbolic Mode* (Princeton: Princeton University Press, 1973),
pp. 354–70; and Lewalski, "Typology and Poetry: A Consideration of Herbert, Vaughan, and

demption function rather in *Upon Appleton House* as indispensable units of a broader cultural discourse of political agency and reformation. Taking advantage of the urgent theological controversies surrounding the key events in Christian history, Marvell explores the problem of the reformation and change of the existing social order. He uses the scriptural figure of the fall from grace to represent the recent national crisis of the Civil Wars, the revolution he depicts as the "wasting" of the "*Paradise* of four Seas" (321–28).[41] Given that the English garden has "fallen," what remains for the poem to explore, within the terms of Marvell's politicotheological discourse, is the problem of "redemption," that action or set of actions that will somehow recover the lost natural—and, by extension, political—order.

The most striking, and certainly the most elaborate, historiographical lesson to "hit" the speaker in his naturalized study concerns itself with just this problem of a political redemption. Like the earlier vision of the trees on the Nunappleton estate, the episode of the oak and the woodpecker highlights a naturalist, vitalist rewriting of the Civil War struggle to purge a corrupt political order:

> But most the *Hewel's* wonders are,
> Who here has the *Holt-felsters* care.
> He walks still upright from the Root,
> Meas'ring the Timber with his Foot;
> And all the way, to keep it clean,
> Doth from the Bark the Wood-moths glean.
> He, with his Beak, examines well
> Which fit to stand and which to fell.
>
> The good he numbers up, and hacks;
> As if he mark'd them with the Ax.
> But where he, tinkling with his Beak,
> Does find the hollow Oak to speak,
> That for his building he designs,
> And through the tainted Side he mines.
> Who could have thought the *tallest Oak*
> Should fall by such a *feeble Strok'*!
>
> (537–52)

Marvell," in *Illustrious Evidence: Approaches to English Literature of the Early Seventeenth Century*, ed. Earl Miner (Berkeley: University of California Press, 1975), pp. 63–69.

41. James Turner has provided examples of the midcentury use of the image of the Fall to suggest contemporary political upheaval, in *The Politics of Landscape: Rural Scenery and Society in English Poetry, 1630–1660* (Cambridge: Harvard University Press, 1979), p. 91.

Critics have for some time identified this passage as Marvell's oblique rendering of the political event that seems most deeply to have affected him, the execution of Charles Stuart, the Royal Oak.[42] Indeed, it could be argued that one of the central images of the "Horatian Ode," the execution of Charles I on the "tragick scaffold," generated for Marvell a tropological structure that appears in numerous guises throughout his work. There can be little question that this passage in *Upon Appleton House* resounds with some of the clamor of the English Revolution. But I propose that we articulate more precisely the particular relation this narrative strikes with that crucial event of 1649. We can discern Marvell's nod to the actual event of the beheading not in the description of the oak but in a brief simile that brings together the activity of the hewel and that of the English regicides: the hewel hacks the trees "As if he mark'd them with the Ax." I would argue that it is in opposition to this embedded image of the ax and its evocation of the decidedly human act of regicide that Marvell arrives at his narrative of the hewel and the oak; it is in contradistinction to the idea of deliberate human action that he develops a fable of the processes of natural selection in the judicious activity of the hewel's tinkling beak. The story of the oak gestures toward the beheading of Charles, but this gesture is dialectical, as Marvell limns a vision of an alternative regicide that functions not as the result of purposive human action but as a link in the causal chain of a natural course of events.

We can perhaps see more clearly the historical dialectic Marvell is constructing if we juxtapose the hewel story with the evocation of the regicide we have already observed, the narrator's sight of a row of ancient trees, "Of whom . . . many fell in war." Both modes of arboreal destruction—the soldiers' axing of the trees in the wood and the hewel's gradual pecking of the giant oak—lead to the same result, the clearing of the forest. But if we consider these passages as models of woodland reformation, and, by extension, as models of any type of social or political reformation, they clearly affirm incompatible positions on the agency behind such historical change: the first passage suggests the willful and plainly human alteration of a given state by organized human effort,

---

42. See Allen, *Image and Meaning*, p. 218; and Harold Toliver, *Marvell's Ironic Vision* (New Haven: Yale University Press, 1965), p. 122. In the poem "All Things Decay and Die" (*The Complete Poetry of Robert Herrick*, ed. J. Max Patrick [New York: Norton, 1968], p. 34), Herrick presses this association of oak and monarch in a manner that might have inspired Marvell's counterfactual historical "allegory":

> That Timber tall, which three-score *lusters* stood
> The proud *Dictator* of the State-like wood:
> I meane (the Soveraigne of all Plants) the Oke
> Droops, dies, and falls without the cleavers stroke.
>   (3–6)

whereas the second, far more elaborate fable imagines this clearing as the exercise of a justice that operates on principles internal to the natural world. It has been argued that Marvell attributes the hewel's destruction of the oak to the work of "Providence,"[43] but such a broad pronouncement cannot properly accommodate the specifically vitalist form Marvell's providence assumes in this scene. Marvell suggests in his utopian vision of historical reformation that it can only be the corrupt subject itself that invites its own destruction. The oak would not have fallen, the narrator insists, had it not already been internally corrupt, "had the Tree not fed / A *Traitor-Worm*, within it bred" (553–54). Trees in this idealized forest participate in the inevitable forces of a natural history: the solid oaks find their proper end as they shoot to heaven, while the corrupt trees seem "to fall content," in gracious recognition of their own corruption.

This question here of the agency behind change, a question that appears curiously to have invited a meditation on trees, exercised many of Marvell's contemporaries during the revolutionary period. We might best come to an understanding of the complex philosophy of trees charging the revolutionary monism of Marvell, and of Winstanley, once we place it in the context of a more explicitly theoretical seventeenth-century consideration of agency. The Cambridge Platonist Ralph Cudworth, writing in the 1650s, follows Aristotle in singling out the example of the felling of a tree in an important discussion of the philosophical problem of causal agency. For both Cudworth and Aristotle, it simply makes no sense to describe a complex event like the felling of a tree without appealing to some form of intentional and external agent, a *"Final, Intending* and *Directive Causality."* Aristotle, explains Cudworth, "ingeniously exposes the Ridiculousness" of the failure to attribute intentional agency to such an event, "telling us,"

> That it is just as if a Carpenter, Joyner or Carver should give this accompt, as the only Satisfactory, of any Artificial Fabrick or Piece of Carved Imagery . . . *that because the Instruments, Axes and Hatchets, Plains and Chissels, happened to fall so and so upon the Timber, cutting it here and there, that therefore it was hollow in one place, and plain in another, and the like, and by that means the whole came to be of such a Form.*[44]

---

43. John Klause, *The Unfortunate Fall: Theodicy and the Moral Imagination of Andrew Marvell* (Hamden, Conn.: Archon, 1983), p. 109. Cromwell had, in fact, proclaimed that the regicides executed Charles I because "providence and necessity had cast them upon it" (*Writings and Speeches of Oliver Cromwell*, ed. W. C. Abbott [Cambridge: Harvard University Press, 1947], 4:473). And it would seem to be in opposition to Cromwell's eager theocentrism that Marvell portrays the execution of the oak.

44. Ralph Cudworth, "The Plastick Life of Nature," in *The Cambridge Platonists*, ed. C. A.

One would be foolish, Aristotle insists, to account for the fashioning of a wooden table by recourse to nothing but the actual physical condition of the standing tree from which the table was made; the *"Intending"* cause of the table must naturally be the carpenter, who deliberately used his tools to square and hew the green tree into a usable piece of furniture.

In juxtaposing Cudworth's defense of Aristotle with the episode of the "tallest Oak" in *Upon Appleton House*, we can discern the very pointed way in which Marvell's narrative thwarts any common-sense view of agency and of historical change. As an allegory of historical reformation—of the gradual clearing away of corruption—the fable of the hewel is perhaps most striking in its failure to attribute the oak's fall to any recognizable *"Intending and Directive Causality."* The consequences of the hewel's actions are utterly inadvertent; although the hewel fells a corrupted oak of tremendous size, there is nothing approaching a deliberate purpose or intention behind the mechanical motions of its beak. Marvell's tree meets its end, *pace* Aristotle, for the simple reason that "it was hollow in one place, and plain in another." The hewel does not destroy the tree so much as it simply cooperates in the tree's own self-determined demise. We saw in Chapter 1 how in 1651 Harvey published a text that reconfigured the facts of the regicide emblematized in his own theory of a decentralized circulation: the doctor staged a scene, in describing his examination of the Viscount Montgomery's wound, of Charles's contented acquiescence to the implicitly republican theory of the heart's insensitivity. In his country-house poem written perhaps in the same year, Marvell, I think, fashions a related narrative of political wish-fulfillment. In a fantasy alternative to the real facts of the English Revolution, Marvell's fable of the oak and hewel imagines a bloodless overthrow of the state, as if Fairfax, in his hewel-like retirement, devoted solely to the values of the household, had without the least trace of aggression or violence purged England of its corrupt monarchy.

It might be fruitful to juxtapose this arboreal narrative of a bloodless revolution with the more theoretical formulation of political change that Marvell would offer twenty years later in his political satire, *The Rehearsal Transpros'd*. In one of that work's most famous passages, Marvell reconsiders the Good Old Cause of the Parliamentarians in the light of a political philosophy that we can identify as revolutionary vitalism: "but upon considering all, I think the Cause was too good to have been fought for. . . . For men may spare their pains where Nature is at work, and the world will not go the faster for our driving. Even as his present Majesties happy Restauration did

---

Patrides (Cambridge: Cambridge University Press, 1969), pp. 290–91. Cudworth is quoting Aristotle's *De partibus animalium*, I, i.

it self, so all things else happen in their best and proper time, without any need of our officiousness."[45] It is most commonly suggested that this expression of political passivity is the result of Marvell's disappointment after the Restoration.[46] But what is perhaps most striking about the curious providentialism in this passage is his stated reliance not on God but on "Nature." Marvell's explicitly political verse often identifies the source of providential control with the traditionally anthropomorphic deity of the Old Testament: it is almost the arbitrary God of Calvinist theology that impels Cromwell's power in the "Horatian Ode" and "The First Anniversary." But Marvell's claim in *The Rehearsal Transpros'd* is rather that men may spare their pains where a providential *nature* is at work. His Restoration critique of political action enlists, if only rhetorically, the faith in an organic revolution embraced by a figure such as Winstanley at the Vitalist Moment. In both the vitalist poems of the Fairfax period and this retrospective vitalism in the Restoration satire, "Nature" functions for Marvell as a self-regulating agent of social change, a force of history that permits a nation to reform itself without "any need of our officiousness," without any need of the deliberate intervention of man with ax.

"The strangenesse of the action"

In the country-house poem's idealist representation of his retreat to his patron's wood, Marvell, as we have seen, seems to validate that mode of passive political aspiration, embraced most famously by Winstanley, that concedes political change to the forces of natural development and growth. Entering the "yet green, yet growing Ark" of Fairfax's wood (an ark superior to Noah's, fashioned without the squaring and hewing of trees), the speaker, retreating from the flooded fields, identifies his submergence in natural process with Noah's redemptive mission, hoping to purify a world corrupted by war. I want here to turn to Marvell's Mower poems, because the midcentury crisis in the conceptualization of revolutionary agency lends a special urgency to his representation of his lovelorn Mower. If in *Upon Appleton House* the retreat into Fairfax's wood constitutes Marvell's attempt, however transitory, to embrace the green idealism of a passive revolutionary

45. Andrew Marvell, *The Rehearsal Transpros'd and The Rehearsal Transpros'd. The Second Part* (Oxford: Oxford University Press, 1971), p. 135.

46. Christopher Hill has suggested, in an argument that complements my own revision of the standard reading of this passage, that Marvell here is expressing a "Harringtonian" view of the inevitable forces of historical change. See *The Experience of Defeat: Milton and Some Contemporaries* (New York: Penguin, 1985), p. 250.

such as Winstanley, then the Mower lyrics surely represent the poet's more distanced, ironic reflection on the viability of a vitalist politics. Marvell lingers in the country-house poem over the exciting possibility of the participation of mankind and nature in a vitalist revolution; the Mower lyrics, however, persistently stage a troubled recognition that such a revolution is unachievable.

It is not my intention to propose the Diggers themselves as the historical referent lurking behind the Mower lyrics, just as it is not my desire to flatten Winstanley's politically motivated treatises to a species simply of literary art. But although Marvell's representations of the Mower may not have the Diggers themselves as a specific or even oblique referent, they nonetheless exploit, for literary purposes, the logical and rhetorical conflicts besetting the texts of the Diggers, or other Digger-like revolutionaries. The tension inherent in Winstanley's prose between human and organic agents of revolution resurfaces, in a modified, highly literary key, in the Marvellian pastoral. The rhetorically complex action of "digging" in Winstanley's texts provides, I believe, an ideologically loaded precipitant for the curiously analogous activity of "mowing" in the lyrics of Marvell, Damon's scythe falling alongside Winstanley's spade as an uncertain, perhaps self-defeating, instrument of change.

An organic immersion in a vitalist world of self-moving matter constitutes the Mower's prevailing ideal in "Damon the Mower," "The Mower's Song," and "The Mower to the Glo-worms." But these shorter lyrics differ crucially from Marvell's *Upon Appleton House* by situating this ecstatic organicism in an irretrievable past. In "Damon the Mower," Marvell relates the baroque giddiness of the Mower's earlier tie to nature by reciting the song Damon used to sing:

> On me the Morn her dew distills
> Before her darling Daffadils.
> And, if at Noon my toil me heat,
> The Sun himself licks off my Sweat.
> While, going home, the Ev'ning sweet
> In cowslip-water bathes my feet.
>   (43–48)

Having sprinkled Damon's body with dew in the morning, Nature used to accept that moisture back again, by noon, in the transmuted form of ingestible sweat. Damon's representation of this delectable baptism moves far beyond, surely, the static images of man's harmonious relation to nature bequeathed to Marvell by the pastoral tradition of Theocritus and Virgil.

Despite Marvell's sensual evocation here of the cycle of evaporation, the Mower lyrics continually recall a pastoral harmony with nature that is less cyclical or even seasonal than linear and progressive. Like the glow-worms before Damon's impassioned fall for Juliana, the dew of "evening sweet" in these lines directs the speaker to the projected end of his journey, home, and as a specifically teleological force, can be seen to function as an organic, vitalist agent of historical change.

The remembered commingling of body and earth assumes the force in all the Mower lyrics, as it does in Winstanley, of natural providential process. In "The Mower's Song," Damon remembers his union with nature as a teleological phenomenon, for his mind had been fully attuned to the immanent providence of vegetable growth:

> My Mind was once the true survey
> Of all these Medows fresh and gay;
> And in the greenness of the Grass
> Did see its Hopes as in a Glass.
> (1–4)

Damon longs here for a return to that union of mind and nature that, we have seen, subtended the period's millenarian hopes in an organically motivated revolution. Like many of the vitalist figurations in Marvell's poetry, the Mower's reading of his "Hopes" in the "greenness of the Grass" is authorized by that scriptural text most frequently cited by the vitalist radicals of the English Revolution. The biblical authority on which the monistic and mortalist revolutionaries relied for the anti-Calvinist vision of a gradual, organically determined world-change is a passage from Paul's Epistle to the Romans, a text that, as we see in Chapter 4, continues to exercise its radical effect in Milton's *Paradise Lost*:[47]

> For I reckon that the sufferings of this present time are not worthy to be compared with the glory which shall be revealed in us. For the earnest expectation of the creature waiteth for the manifestation of the sons of God. For the creature was made subject to vanity, not willingly, but by reason of him

47. Milton cites these verses to defend his mortalist heresy and his materialist millenarianism in *De doctrina*, in *Complete Prose Works* 6:399, 6:450, 6:496, 6:633. See Overton's use of this passage in his early, influential mortalist treatise, *Mans Mortalitie* (1644), p. 70. Winstanley echoes this text throughout his writings but most distinctly and repeatedly in *The New Law of Righteousness* (1649), pp. 149–246 of *Works of Winstanley*. Geoffrey Hartman, "Marvell, St. Paul, and the Body of Hope," in *Beyond Formalism: Literary Essays, 1958–1970* (New Haven: Yale University Press, 1970), pp. 166–69, argues for the importance of this Pauline text for the Mower lyrics.

who hath subjected the same in hope; because the creature itself also shall be
delivered from the bondage of corruption into the glorious liberty of the children of God. For we know that the whole creation has been groaning in labor
pain until now; and not only the creation, but we ourselves, who have the first
fruits of the Spirit, groan inwardly while we wait for adoption, the redemption
of our bodies. (Rom. 8:18–23)

With the key word *ktiseos*, translated properly by the King James as both
"creature" and "creation," Paul sheds his Hellenic dualism for a visionary
Hebraic monism, conglobing into a single mass of sentient expectation the
bodies of earth and earthling. In this remarkable vision that conjoins creature
and creation in one harmonious act of labor and groaning, Paul provides a
wide array of midcentury monists, mortalists, and hermeticists with a powerful sanction for their dream that the labor required to institute the millennium be not human labor only but the labor also of the animal and
vegetative forces of nature. On the authority of this particular prophecy of
the deliverance from bondage not only of the children of God but of the
entire animate creation, the hope in millennial reformation could be located
in the greenness of the grass, the growing, groaning powers of the body of
creation.

The Pauline image of a groaning that is no more human than earthly
doubtless proved attractive to the period's passive revolutionaries because of
its elegant vitalist reworking of the most common figure for political agency,
masculine human labor. For Paul here, it is not men's labor, or the work of
direct political intervention, that will fulfill the "earnest expectation" of the
"manifestation of the sons of God." The groaning labor of revolution here
is less masculine than feminine, less human than terrestrial, as the "whole
creation," "groaning in labor pain," struggles to deliver humanity in an apocalyptic parturition that Paul names "the redemption of our bodies." Paul's
unusually corporeal image of the promised end of Christian history is recast
by the revolutionaries of the Vitalist Moment, as it is by Marvell's Mower,
in the most literal fashion imaginable: the final union is figured as the actual
physiological convergence of creaturely and vegetative bodies. But the passive acquiescence to a sentient, laboring creation characterizes, in the Mower
lyrics, only Damon's former relation to the natural world. We must turn to
Marvell's representation of the Mower's present action to understand the
particular way in which these poems offer a critical reflection on an organicist
ideal we can identify, roughly, as Pauline and Winstanleyan.

Like *Upon Appleton House*, the Mower lyrics evince a politically resonant
dialectic of growth and arrest that hinges on a narrative element roughly

assimilable to the founding crisis of human history, the Fall. Like the Civil War, which is seen in *Upon Appleton House* to cause the lapse of an entire nation, the violent eruption of the Mower's passion for Juliana, in "Damon the Mower," "The Mower to the Glo-Worms," and "The Mower's Song," prompts his tragic disseveration from the earth's immanent vitalist progressivism. Having identified Marvell's imposition of the structure of Christian history onto his notion of the providential development of nature, we can hear the revolutionary reverberations of the rhetorical gesture that constitutes the closure of these poems. Marvell's Mower acts with his scythe to redeem his former state, attempting to recover, by means of human labor, a Pauline coalescence with a laboring creation.

In "The Mower's Song," Damon assumes a prophetic voice when in his frenzy he vows to avenge himself on the unresponsive world around him with an act of violent, apocalyptic mowing: "And Flow'rs, and Grass, and I and all, / Will in one common Ruine fall" (21–22). In "Damon the Mower," Marvell describes in apocalyptic tones the Mower who in his haste cuts his own ankle with his scythe:

> While thus he threw his Elbow round,
> Depopulating all the Ground,
> And, with his whistling Sythe, does cut
> Each stroke between the Earth and Root,
> The edged Stele by careless chance
> Did into his own Ankle glance;
> And there among the Grass fell down,
> By his own Sythe, the Mower mown.
>
> (73–80)

When we consider the literal plunge into grass the Mower takes in these lines, we might say, from one perspective, that he has succeeded in his attempt to regain his paradisal conjunction with vegetable providence. But the emphasis in these poems on Damon's tragic folly pressures us to acknowledge the ironic self-defeat of this attempt—perhaps any attempt—to force an alliance with the forces of nature. As Marvell's repeated use of images of *falling* suggests, an attempted act of redemption can only recapitulate an originary act of destruction: the cause of radical redemption is too good to fight for. Embodying within their structure the most defeatist component of Marvell's own *Rehearsal Transpros'd*, these lyrics look even further ahead to the dispiriting wisdom of Nietzsche's reading of the natural salvation envisioned by St. Paul: "we feel . . . we are not the men toward whom all Nature presses for her redemption."[48]

It is in reference to the vexed status of the concluding action of these poems that we can locate the most prominent node of intersection between Marvellian lyric and Winstanleyan treatise. Surely the dominant irony behind Marvell's use of the mower figure resides in the fact that, even as Damon idealizes the passive reciprocity he once enjoyed with his physical surroundings, his task as mower requires a systematic exercise of violence on that natural world. Damon may, in "The Mower's Song," see his hopes in the greenness of the grass, but as Marvell reminds us in "Damon the Mower," it is Damon's terrible duty to cut down that symbol of hope: "Sharp like his Sythe his Sorrow was, / And wither'd like his Hopes the Grass" (7–8). It is its role as self-punishing instrument of redemption that suggests the alliance of Damon's scythe to Winstanley's spade. Winstanley may have foretold, in treatise after treatise, an autochthonous utopia that would spring from the soil like a tree, but the action available to him was the unfortunate action of digging, the systematic removal of trees—the very symbols of hope—from the commons.

The irony of this particular form of symbolic political action does not appear to have been lost on Winstanley's contemporaries. In his visit to St. George's Hill on May 26, 1650, Fairfax had attempted to divine the significance of the Diggers' activity. But this pursuit of political intention was no less difficult for Fairfax, I suggest, than it is for us. An anonymous pamphlet published shortly after the encounter reported that the Diggers, in response to Fairfax's queries, "gave little satisfaction (if any at all) in regard of the strangenesse of the action."[49] Winstanley, it must be said, was not himself blind to the "strangenesse of the action" of his digging, an action that hovered uncomfortably between hesitant political intervention and aggressive political passivity. Echoing as so often the Pauline vision in Romans, he denounces political intervention as not only ineffective but possibly even harmful to the cause: "The whole earth we see is corrupt, and it cannot be purged by the hand of creatures, for all creatures lie under the curse, and groan to be delivered, and the more they strive, the more they entangle themselves in the mud."[50] Like Marvell, Winstanley follows Paul into a state of millenarian resignation: those revolutionaries who, subject to "hope," attempt to precipitate the glorious end themselves are, in Paul's words, "sub-

---

48. Friedrich Nietzsche, *Unmodern Observations*, ed. William Arrowsmith (New Haven: Yale University Press, 1990), p. 195.

49. *The Speeches of the Lord General Fairfax, and the Officers of the Armie to the Diggers* (London, 1649), sig. E4ᵛ; quoted in Corns, *Uncloistered Virtue*, p. 166.

50. Winstanley, *New Law of Righteousness*, in *Works of Winstanley*, pp. 186–87.

ject to vanity." Marvell's Mower and Winstanley's Digger are doomed by profession to careers of strange, ironic action. The rhetoric that begins with a palpable anticipation of realizable change ends in a self-induced defeat: the Mower mown, the Digger dug.

In the "Horatian Ode," Marvell can claim that the "bleeding Head" (69) of Charles augurs the success of the new state, and in *Upon Appleton House*, as we see in the next chapter, the sacrifice of Mary Fairfax to the priestly knife in marriage promises "some universal good." But the Mower lyrics, like many sections of the country-house poem, follow a Winstanleyan resignation to nature and question the faith in the redemptive potential of such sacrificial acts. What is perhaps most central to Marvell's pastoral historiography is the way in which these dramatic coups de grace not only fail but actually preclude all hope in an ideal future. Even though the intended goal of the poems' final actions may be to speed the process of man's reabsorption into nature, the mere fact of human assertion in these lyrics seems to kill the possibility of an organic, Winstanleyan apocalypse. Like the speaker of *Upon Appleton House*, as he importunes the "courteous *Briars*" to subject him to a woodland crucifixion (609–16), the Mower is placed in the paradoxical position of actively forcing himself into a passive relationship with nature. This vitalist revolutionary can but lapse into sybaritic buffoonery as he embraces the body of earth with too much erotic intensity.

It is the grim irony accompanying the hope in any human action that motivates Marvell's use of the georgic figure of the Mower. Annabel Patterson and Anthony Low, extending Raymond Williams's analysis of the political content of the first-generation country-house poems, have each argued that Marvell's innovative use of a mower bespeaks an interest in what Patterson calls the "question of land-ownership and rural labor."[51] Rosemary Kegl has argued, further, that the Mower lyrics are to be judged complicit in the socioeconomic oppression of agricultural workers like Damon.[52] These readers are, of course, right to claim that the Marvellian pastoral is entangled in the complexities of its historical moment. But I think we have seen that the historical referent of these poems is less a material set of labor conditions

51. Patterson, "Pastoral versus Georgic," p. 262. Low, *Georgic Revolution*, given to the unlikely but happy thought of Marvell's revolutionary sympathy with agrarian labor, is constrained to produce a positive reading of the Mower's leveling actions in *Upon Appleton House*: "A mower, glancing down to find his scythe edged with innocent blood, may well blanch, but if he refuses to press on until he has won the victory, that bloodshed will have been in vain" (p. 287). A historically informed, emblematic reading of the Mower figure, closer to the interpretive scheme I have sketched here, can be found in Leah S. Marcus, *The Politics of Mirth: Jonson, Herrick, Milton, Marvell, and the Defense of Old Holiday Pastimes* (Chicago: University of Chicago Press, 1986), pp. 233–36, 250–52.

52. Kegl, " 'Joyning my Labour to my Pain,' " pp. 89–118.

than a culturally valent discursive system that employs "labor" as a figure for political activity. Marvell's poems exhibit less interest in the socioeconomic status of agrarian labor than we find in his predecessors in the country-house genre, Jonson, Carew, and Herrick. Marvell, rather, deploys the Virgilian dialectic of pastoral and georgic as a symbolic system with which to explore what was for him the more pressing—albeit more abstract—question of the dynamics of political agency. The georgic action of mowing, like the action of digging, may be from a practical standpoint necessary, a beneficent activity intended to provide fodder for animals. But we need not imagine that Marvell has attempted to construct in these lyrics an agriculturally viable image of the production of hay: the representation of a mower who, in the violence of his frenzy, shows an unlikely tendency to "cut / Each stroke between the Earth and Root" ("Damon the Mower," 75–76) does not, I submit, point to a recognizable practice of graminiculture. Surely the capitalized letters bestowed on "Earth" and "Root" in this stanza signal the emblematic quality of Damon's activity, evoking, I suggest, the many mid-seventeenth-century attempts at political and cultural deracination.

Damon's zealous stroke here may glance, as we have seen, at the radical agenda of a Winstanley, a figure who frequently imaged his political goal as the uprooting of an entire social system. But many of the period's radical movements described their action as deracination, as the disruption of the "ground" or foundation of society. The image of Damon's scythe evokes the general principle of revolution, suggesting the riots of the 1640s attending the Root and Branch Petition or the radical Puritan desire, forwarded by the Hermeticist Roger Crab in the 1650s, to deploy "The English Hermites Spade at the Ground and Root of Idolatry."[53] In the highly emblematic world of Marvell's lyrics, the Mower's intervention in the realm of natural process—his "stroke between the Earth and Root"—functions as a gesture of revolutionary activism that invariably fails. The inevitable self-destruction that for Marvell accompanies any human action underscores the dispiriting, baffling political wisdom that man should spare his pains where nature is at work.

The paradox of revolutionary agency that besets the texts of Marvell and Winstanley is clearly not unique to these two writers of the 1650s. Though with varying degrees of political and theological resonance, the paradox might even be endemic to all intellectuals struggling with the question of agency at the Vitalist Moment. This period, as we have seen, witnesses a tremendous explosion in natural philosophical speculation, the emergence of

---

53. Roger Crab, *The English Hermite and Dagons-Downfall* (1655, 1657), ed. Andrew Hopton (London: Aporia, 1990), p. 28.

an entire discipline—philosophical alchemy—devoted to a belief in the regenerative, self-moving powers of natural, organic substances. The conflict
in the Mower poems between Damon's passive idealism and his awkwardly
aggressive attempts to realize that passivity presents in an acute form the
crisis of agency confronting not only Winstanley but an array of radical voices
in mid-seventeenth-century England. Marvell's pastoral poems struggle with
one of the most troubling questions his century posed: what is the point of
political action in the face of a revolution overseen and perhaps even controlled by a higher, inhuman power? His lyrics transmute, through a literary
alchemy, the logical problems attending this urgent question into the irony
and paradox we have been taught to find constitutive of his verse. Marvell's
poems at once embrace and dismiss the Winstanleyan faith in a laboring
creation, finding, and losing, the seeds of revolution in what *Upon Appleton
House* names the "unfathomable Grass" (370).

3 Marvell and the Action of Virginity

> Nothing is superior to chastity in its power to restore mankind to Paradise.
> —THEOPATRA, in Methodius, *Symposium: A Treatise on Chastity*

> Virginity is the name of whatever is capable of being absolutely lost.
> —ANGUS FLETCHER, *The Transcendental Masque*

The participatory labor of creature and creation envisioned by Paul formed one of the bases, as we have seen, of a mid-seventeenth-century fantasy of a passive revolution. The Pauline doctrine in Romans 8 of a bodily redemption occasioned by a creation "groaning in labor pain" was felt to authorize a theological heresy that at least some figures in the 1650s held accountable for the passions that led to civil war.[1] We have examined, in Chapter 2, how Marvell uses *Upon Appleton House* to explore the attractions of this visionary alternative to the shooting and killing, the squaring and hewing endemic to a revolution that, even after the establishment of the republican Commonwealth in 1649, still possessed a propensity for violence. We need still to explore, however, Marvell's investment in an important, if curious, corollary to the doctrine of the passive revolution. I argue here that the attraction not only for Marvell but for a few of his contemporaries of the Pauline vision of natural redemption can be traced in large part to the considerable prestige Paul accorded the bodily condition of virginity.

The pastoral ideal of many of Marvell's poems is unquestionably an ideal of sexual abstinence, the paradisal retreat envisioned in "The Garden" as the "happy Garden-state, / While Man there walk'd without a Mate." Poems such as "The Garden," the Mower lyrics, and *Upon Appleton House* consis-

---

1. Joseph Frank, in *The Levellers: A History of the Writings of Three Seventeenth-Century Social Democrats* (Cambridge: Harvard University Press, 1955), cites articles from two Royalist newspapers of 1648 that "complained that the Mortalist Heresy was one of the more subversive ideas engendered by the Civil War, and that it was partly responsible for the revolutionary nature of that conflict" (p. 299). The best account of the mortalist heresy is Norman T. Burns, *Christian Mortalism from Tyndale to Milton* (Cambridge: Harvard University Press, 1972).

tently align the delicious passivity of their speakers with a retreat from sexual contact, voicing everywhere a preference for the unitary subject's languorous solitude over the feverish work of sexual assertion. The virtuous withdrawal from sexual experience becomes, perhaps, the central component of the *vita contemplativa* envisioned in the Marvellian pastoral. It is no doubt in relation to this ideal of a virginal retirement that Marvell forges a figural alliance between sexual and political aggressivity. His exercise in carpe diem, "To His Coy Mistress," points us most clearly to the complex, almost inextricable connection in many of his works between the rough strife of the *vita activa* and the vital activity of sexual intercourse. The spheres of action and retirement described in Marvell's poetry encompass more, however, than merely ethical or sexual alternatives. The conventional images of the active and contemplative lives function for Marvell as discursive units indispensable for the larger analysis of revolutionary agency; they figure, in particular, as distinct conceptual models for the institution, sought by so many in the revolutionary period, of the millennial kingdom.

The Fifth Monarchists and other radical sects of the mid-seventeenth century relied heavily on a faith in the efficacy of militaristic attempts to hasten the onset of the Last Days. Other, more pacific minds maintained, however, that God would respond not to man's direct action but to a more passive exercise of morality, the kind of virtue fostered, for Winstanley, on the commons of St. George's Hill, for Marvell, in the gardens of Fairfax's estate. In situating Marvell's country-house poem amid some of the theological discourses of the Vitalist Moment, I want to trace the poet's transposition of the period's millenarian thought onto the social alternatives of virginity and marriage. Marvell, as we will see, appropriates the sexual politics of some of the period's most radical vitalists to assert the political, even millenarian, implications of the passive retirement of the innocent virgin.

"The faire flowre of Chastity and vertue virginall"

However heartfelt we may have found Marvell's passivist rewriting of the revolution in *Upon Appleton House*, it is not a vision of Winstanleyan organic process, or resigned expectation, with which Marvell concludes that poem. Ultimately, he seems to reject in the country-house poem the imagined efficacy of the passive revolution. I want to begin here with a consideration of the crucial moment near the end of *Upon Appleton House* in which Marvell represents the entrance into the poem of his patron's daughter, Mary Fairfax. Just as the speaker has almost fully dissolved into Winstanleyan union with the natural world—"Stretcht as a Bank unto the Tide" (644)—he grows conscious that "The *young Maria* walks tonight" (651).[2] As his pupil proceeds through the landscape, the narrator's monist fantasy of the redemption of the creation's body evaporates abruptly in the girl's numinous presence. With a preternatural power like that of Marvell's Cromwell, Marvell's pupil seems able to "contract the scattered force of time," concentrating a supernatural energy that clearly transcends the mundane operations of natural process:

> *Maria* such, and so doth hush
> The *World*, and through the *Ev'ning* rush.
> No new-born *Comet* such a Train
> Draws through the Skie, nor Star new-slain.
> For streight those giddy Rockets fail,
> Which from the putrid Earth exhale,
> But by her *Flames*, in *Heaven* try'd,
> *Nature* is wholly *vitrifi'd*.
>
> (681–88)

Marvell charges his description here with the rhetoric of divine intervention and extraordinary providential control, taking pains to distinguish Maria's prodigious powers, "in *Heaven* try'd," from those less exalted flames that "from the putrid Earth exhale." From the moral perspective introduced at Maria's entrance, that idealized fusion of man and nature is dismissed as delusory and self-indulgent. Whereas the narrator had participated in a natural world that seemed to move by a gradual, terrestrial progression toward a final heavenly end, Maria's entrance introduces to the poem what is perhaps a more orthodox Puritan understanding of Christian history, a concession to the violent discontinuity of apocalypse: in her presence, "*Nature* is wholly *vitrifi'd*," reduced, as if by the final conflagration, to the crystalline state that had once been thought to resemble "the untainted substance left after the earth has been tried by fire."[3]

In view of what is surely the inappropriate application of apocalyptic rhetoric to Fairfax's young daughter, it is not unreasonable for us to ask on

2. *The Poems and Letters of Andrew Marvell*, 2 vols., ed. H. M. Margoliouth, rev. Pierre Legouis (Oxford: Clarendon, 1971), vol. 1. Quotations drawn from this work are cited by line number in the text.

3. Kitty W. Scoular, *Natural Magic* (Oxford: Clarendon, 1965), p. 178. Margaret Cavendish rehearses this thesis: "since the last Art of Chymistry (as I have heard) is the Production of glass, it makes perhaps Chymists believe, that at the last day, when this Word [sic] shall be dissolved with Fire, the Fire will calcine or turn it into Glass: A brittle World indeed!" (*Observations upon Experimental Philosophy*, p. 78).

what basis the speaker invests Maria with this surcharge of divinity. When the narrator asserts Maria's superiority to the other members of her "fond sex," he establishes the source of her powers; it is nothing less than the prospect of an intrusive human action—the action of marriage—that removes Mary Fairfax from the natural world to a realm of unapproachable divinity:

> Hence *She* with Graces more divine
> Supplies beyond her *Sex* the *Line*;
> And, like a *sprig of Misleto*,
> On the *Fairfacian Oak* does grow;
> Whence, for some universal good,
> The *Priest* shall cut the sacred Bud;
> While her *glad Parents* most rejoice,
> And make their *Destiny* their *Choice*.
>
> (737–44)

The final section of *Upon Appleton House* swells in a joyful anticipation of the "universal good" that will follow the marriage of Fairfax's only child. We remember that earlier in the poem, the nuns in the Nunappleton Priory had struggled to delay the triumphs of William Fairfax, the present lord's sixteenth-century ancestor, by sequestering Isabel Thwaites in a morally dubious state of virginal retirement: "against Fate, his Spouse they kept; / And the great Race would intercept" (247–48).[4] Here the current Fairfaxes' choice to accept the destined marriage of their daughter fulfills the apocalyptic prophecy attached to that first heroic Fairfax, the patriarch "whose Offspring fierce / Shall fight through all the *Universe*" (241–42). Marvell's employer, formerly the head of the Parliamentary army, had withdrawn, of course, from political life to his family estate at Nunappleton, protesting at once the act of regicide and the prospect of a military invasion of Scotland. But by translating his daughter beyond the walls of virginal isolation to the public world of matrimony, the former Lord General can in some way redeem his own ignominious retreat from his commitment to the public sphere. Marriage functions in this poem as a mode of public action, an interventionist attempt to alter and improve the languorous private state of fallen virginal nature. By heralding the possibility of the continuation of the Fairfacian line, the marriage of Maria can even redeem the nation depicted throughout the poem as "fallen" in war.

---

4. For a rare antithetical reading of Marvell's nuns, focusing properly on the attractiveness of their cloistered position, see Judith Haber, *Pastoral and the Poetics of Self-Contradiction* (Cambridge: Cambridge University Press, 1994), pp. 129–32.

The representation of this redemptive action recalls two auspicious events to which Marvell so often returns, the execution of the Stuart monarch and what is for Marvell that execution's typological precursor, the crucifixion of Christ.[5] The sacrifice of Maria's virginity on the marriage bed retraces in outline the sacrifice of Charles, who in the "Horatian Ode" "bow'd his comely Head / Down, as upon a Bed" (63–64). Like the "bleeding Head" of a king that augured in the "Ode" the "happy Fate" of a newly designed nation (67–72), the disseverment of Maria's maidenhead will by a mysterious process of sacramental causation usher in a "universal good." But the claim for a global melioration also calls on one of marriage's more traditional associations, invoking the alliance of matrimony with the operation of sacrifice and reward that forms the basis of the Christian Redemption.[6] Like the crucifixion of Christ, which functions in Christian history as the violent sacrifice of a child that institutes the redemption of fallen man, Fairfax's sacrifice of his daughter on the altar of marriage will institute the redemption of a fallen nation. Decisive action throughout Marvell's poetry assumes the figural shape of sacrifice—the violation, in this instance, of the private body and the private sphere—and Fairfax's sacrifice here constitutes an action that, if it cannot actually reform "that dear and happy Isle" of England (321), might at least invite a transcendent apocalypse that redeems man by removing him from the temporal world.

Most critics of the poem have accepted without question its sanction of the narrator's jubilant claim that the Fairfaxes have blithely made their daughter's destiny their choice. We should not, however, overlook the degree to which Marvell's image for this redemptive marriage represents an unsettling divergence from the poem's established ethic of natural growth and development. In a poem so fully invested in the ideology of the "green world," Marvell's figure for Mary's marriage—the priest's cutting of the "sa-

---

5. Blair Worden, in "Providence and Politics in Cromwellian England," *Past and Present* 109 (1985): 90, argues that it was an inveterate logic of sacrifice among the Puritan revolutionaries that enabled the commitment to regicide: "The execution of Charles I is inexplicable without reference to the Old Testament 'vengeance' breathed by Cromwell after the second Civil War, and the biblical theory of blood-guilt which was as potent a force among the regicides in January 1649 as it was to be among Cromwell's soldiers in Ireland later in the year."

6. For Aquinas, marriage effects its sacramental impact "through Christ having represented it by His Passion" (*The Summa Theologica of St. Thomas Aquinas*, vol. 3, *Supplement* [New York: Benziger Brothers, 1948], Q. 42, A.3, quoted in James Turner Johnson, *A Society Ordained by God: English Puritan Marriage Doctrine in the First Half of the Seventeenth Century* [Nashville: Abingdon, 1970], p. 44). Marvell's unusually sacramental representation clearly marks a pronounced distance from the contemporary Puritan emphasis on the primacy of companionship in marriage, constituting instead an ironic version of the Catholic association of matrimony with the Christian Redemption. See Johnson, *Society*, p. 44, for a view of the Puritan tendency to see marriage as a covenantal rather than a sacramental act.

cred Bud"—carries a rhetorical weight that nearly overwhelms the speaker's straightforward praise: the sacrifice of the girl's virginity seems to assume the shape of a virgin sacrifice. The image of oak and mistletoe is not new to the poetic praise of marriage: it provides a central figure for one of the first epithalamia written in English, "Let mother earth now deck herself in flowers," from Sidney's *Arcadia*. The shepherd Dicus in that work praises the virtuous husband and wife as "oak and mistletoe, / Her strength from him, his praise from her do grow."[7] Marvell, needless to say, has violated the traditional logic of the epithalamic coupling of these two plants. To image the mistletoe not as wedded to a husband oak but as cut off from the parent oak is to reorient the generic tenor of marriage from romance to tragedy. The priest's druidic disseverance of the delicate sprig of mistletoe evokes inevitably the other tragic actions narrated in the poem: the violation of nature committed by the wanton cutter of trees, or by the leveling cutters of grass in the poem's harvest section.[8] Although Marvell scripts this assertion of the girl's marital prospects in the mode of encomium, the image conveys a disturbing sense that this particular redemption is not without its cost.

We might better understand the suppressed pathos of the Fairfaxes' resignation to destiny if we place Marvell's image of the cut bud, the figure in *Upon Appleton House* for Mary's marriage, alongside its later appearance in Marvell's poetry, in the 1658 elegy to Cromwell. The Lord Protector's death had followed by a month the death of his daughter, Eliza. In "A Poem upon the Death of O. C.," Marvell employs the image of a vine severed from the family tree to signify not the marriage but the death of the daughter. More dramatically, the cutting of the bud from the Cromwellian oak does not result in her father's joy but leads, at least according to Marvell, to the father's death:

> If some deare branch where it extends its life
> Chance to be prun'd by an untimely knife
> The Parent-Tree unto the Griefe succeeds,
> And through the Wound its vital humour bleeds.
>
> (93–96)

By lingering with such care over Cromwell's consuming grief, Marvell presents in the elegy not simply a normative parental response to a child's tragic

7. Sir Philip Sidney, *The Countess of Pembroke's Arcadia*, ed. Maurice Evans (Harmondsworth: Penguin, 1977), p. 693.

8. Robert Cummings, in "The Forest Sequence in Marvell's *Upon Appleton House*: The Imaginative Contexts of a Poetic Episode," *Huntington Library Quarterly* 47 (1984): 192–96, argues persuasively that it is the seventeenth-century understanding of Druidic sacrifice that informs the description of Mary's marriage.

destiny but the far more unsettling notion that this private loss will precipitate the public tragedy of the downfall of an entire government, the "Parent-Tree" (no less Cromwell than all England) bleeding its vital humor through the wound. These lines, which deploy a metaphor of cutting quite explicitly to figure the untimely demise of a child, help us, I think, uncover the hidden intimations of mortality borne by this image of the violation of the integrity of the "*Fairfacian Oak*."

Behind the vision of the "*Fairfacian Oak*" in Marvell's description of his pupil's future marriage stands the other oak we have observed in *Upon Appleton House*. And it is with the image of the oak tree that Marvell draws the idea of Mary's future marriage into relation with the fable of the hewel we examined in Chapter 2, marking the priest and the hewel as distinct types of historical agents and the two narratives themselves as distinct paradigms of revolutionary change. I want now to explore how these narratives of the Royal Oak and the Fairfacian Oak gesture to the contemporary concern we have noted with the extirpation of trees and, by emblematic extension, with the culture's broader critical reflection on the problem of political agency. The two oak-tree narratives in Marvell's poem point not just to his interest in the razing of trees by a group such as the Diggers; they point especially, I propose, to Marvell's canny engagement of a politically valent fable told and retold during the years of the Civil Wars and Interregnum. From James Howell's *Dodona's Grove: or the Vocall Forrest* (1640) to the fable "Of the Husband-man and the Wood" in John Ogilby's *Fables of Aesop Paraphras'd in Verse* (1651), stories of oak trees and their tragic fates were deployed, as Patterson has demonstrated, to provide embedded, and usually Royalist, interpretations of contemporary political events.[9] In his *Discourse of Forest Trees* (1664), John Evelyn recounts a particular fable in circulation during the Interregnum that was but one of the many "histories of groves that were violated by wicked men, who came to fatal periods; especially those upon which the misselto grew, than which nothing was reputed more sacred":

> I am told of the disasters which happened to the two men who (not long since) fell'd a goodly tree, call'd the Vicar's Oak, standing at Nor-Wood (not far from Croydon) partly belonging to the arch-bishop, and was limit to four parishes, which met in a point; on this oak grew an extraordinary branch of misselto, which in the time of the sacrilegious usurpers they were wont to cut and sell to an apothecary of London; and though warn'd of the misfortunes observed

---

9. Patterson discusses Ogilby's fable and its roots in Aesop and in Spenser's "February" eclogue in *Fables of Power*, pp. 60–61, 87–89.

to befall those who injured this plant, proceeding not only to cut it quite off, without leaving a sprig remaining, but to demolish and fell the oak it self also: The first soon after lost his eye, and the other brake his leg; as if the Hamadryads had revenged the indignity.[10]

The popular genre of the "histories of groves that were violated by wicked men," a fable responding not only to the general concern with the common disregard of ancient institutions but to the more specific problem of arboreal destruction, supplies an important imaginative source for the representation of both the oaks in *Upon Appleton House*. Like the oak in the fable of the hewel, the Vicar's Oak (suggestive of the ill-fated Archbishop Laud) suggests the vulnerability of any sanctioned authority, in this "time of . . . sacrilegious usurpers," to the deliberate violence of ax-bearing men. Like the *"Fairfacian Oak"* in the stanza on Mary's marriage, this tree is prized for its "extraordinary branch of misselto," which is purposefully "cut . . . quite off."

It is easy to see that the vitalist Marvell, however anti-Royalist we presume him to be, would have found himself sympathetic to the general sentiment of these "histories of groves" concerning the sacredness of organic life. Like Marvell's poem, this folk narrative harbors a reverence for those goodly trees "than which nothing was reputed more sacred," and it measures this sylvan respect against a destructive impulse specific to the period of the Civil Wars and Interregnum. But although the folktale of the oak seems nonetheless to linger behind the plangent narratives of *Upon Appleton House*, it is important to note the pointed distinction between Evelyn's Vicar's Oak and the Fairfacian Oak with which Marvell concludes the action of his country-house poem. Where in Marvell the cutting of the sacred bud of mistletoe is the occasion for communal joy, the excision of mistletoe in the folktale, an act motivated by pecuniary gain, brings unmitigated disaster, as the men in search of the valuable leaf proceed "not only to cut it quite off, without leaving a sprig remaining, but to demolish and fell the oak it self also." It is this purposeless, supererogatory violence to the "oak it self" that constitutes the tragic center of Evelyn's fable. And I think that this grim popular narrative of the demolition of an entire oak can be seen to cast a pall over the related stanza in Marvell's poem.

Despite the joy asserted by the narrator in anticipation of Mary's marriage, the cold, even brutal, image used to figure that marriage—the cutting of the "sacred Bud"—invites a response other than joy and invokes at this critical moment the very differently inflected conclusion of the popular tale: the violation of the poem's commitment to a viridescent revolution, a violation

<hr />

10. Evelyn, *Sylva*, 2:249–50.

implicit in the image of the cutting of the mistletoe, suggests the eerie possibility that this action might result not in a "universal good" but in the tearing up of the sacred oak of England, root and branch. The folktale's shadowy presence just behind Marvell's text provides us with an alternative, tragic conclusion to the otherwise romantic poem, lending a narrative shape to the poem's implicit distrust in the value of even the most sanctioned human action.

The Virginal Politics of Agency and Organization

We have seen how Marvell's disturbing delineation of matrimony raises questions concerning its redemptive value. Mary Fairfax's marriage is figured as severing bud from plant, drawing to a violent close that gradual process toward the "green Age" that forms the poem's vitalist ideal of nonviolent, reformist political change. As a prospective bride, however, Mary is allied with a supernatural providential force that endows her sacrifice with extraordinary meaning: she exudes an energy that not only is celestial in origin but actually fixes the groaning motions of nature in an apocalyptic stasis. But although the rhetoric of the eschatological transcendence of worldly history is logically incompatible with a more Winstanleyan, millenarian hope in a budding reformation here on earth, Mary's appearance in *Upon Appleton House* does not register the poem's unequivocal rejection of the redemptive potential of a natural revolution. The poem has instead inscribed within its extravagant encomium of Mary's nuptial future a complex but identifiable counterargument. In the midst of the poem's final concession to the seemingly inexorable necessities of political action and divine assistance is a subversive counternarrative that questions not only the redemptive value of a sacrificial action such as matrimony but the entire ideological scheme of which that value is a part. This counternarrative, I believe, in amplifying the impact of the simple virtue of virginity, supplies the poem with a radical, decidedly sectarian set of images for the curious, distant ideals of liberal agency and liberal organization.

With Marvell's image of the sacrifice of the "sacred Bud" a considerable gap opens between the poem's argument of praise and the pathos of its imagery. It is precisely at this moment of tonal incongruity that the poem admits the viability of an important alternative to the logic of redemptive sacrifice and apocalyptic transcendence. In the narrative time of the poem, Mary Fairfax has not yet married. We must now turn to those stanzas at the end of the poem in which Marvell indulges in a rhapsodic prophecy of what might happen on the estate *before* the girl's removal from Nunappleton, before the excision of the mistletoe from the Fairfacian Oak. Dissociating the virtue of physical virginity from the taint of papist immorality established in the convent scene, the narrator imagines the historical potential of Mary's premarital condition while "yet She leads her studious Hours" on the estate with her tutor:

> Mean time ye Fields, Springs, Bushes, Flow'rs,
> Where yet She leads her studious Hours,
> (Till Fate her worthily translates,
> And find a *Fairfax* for our *Thwaites*)
> Employ the means you have by Her,
> And in your kind your selves preferr;
> That, as all *Virgins* She precedes,
> So you all *Woods, Streams, Gardens, Meads*.
>
> (745–52)

No critic of this poem has questioned the incompatibility of this progressive, specifically naturalist vision with the narrator's earlier image of Maria's eschatological vitrification of all things earthly and temporal. But Marvell quite clearly maintains a sharp distinction between these two paradigms of Mary Fairfax's relation to future history. As long as Mary retains her virginity, he suggests, there is still some hope for a natural mode of historical progress, one that does not require the flames of heaven to temper the earth but that allows the creation's own natural processes to effect a gradual and far gentler terrestrial apocalypse. Before she is to submit herself to the nuptial altar, Mary has in the "Mean time" the capacity to bestow on the natural world around her the virtues of her virginity.[11] In a remarkable inversion of traditional epithalamic values, Marvell asserts that it is virginity, rather than marriage, that can kindle a fructifying and regenerative force in nature.

In this hyperbolic description of a young girl's magical hold over the natural world, Marvell imagines that the powers exercised by Mary derive from two distinct sources. He maps a dialectic of nature and grace—of a Winstanleyan, millenarian hope in a natural reformation and the eschatological hope in a final, supernatural transcendence—onto the ethical alternatives of virginity and marriage; and underlying this dialectic is an uneasy tension between irreconcilable projections of historical potential. The angels in Milton's *Paradise Lost* will go on to voice their confusion as to whether "New Heaven and earth shall to the ages rise, / Or down from Heaven descend" (10.647–48). But here in Marvell, the period's conflicted allegiance to com-

11. For a different reading of Marvell's phrase "mean time," see Stocker, *Apocalyptic Marvell*, pp. 133–36.

peting modes of cosmic agency is displaced onto the seemingly homier question of an individual's sexual alternatives. The state of virginity promises a future that takes the shape of a natural, incremental reformation rooted firmly in the temporal world. The fate of marriage, however, bodes a starker, more apocalyptic mode of historical change, the absolute transcendence of natural process that results from the divine response to heroic acts of human sacrifice.

It is at this point that we must register the degree to which Marvell departs from the conventional homologizing among the antinomies of marriage and virginity, the active and the contemplative lives, and the natural and supernatural forces of causation. It was a commonplace in the period, as Marvell's contemporary Richard Baxter later explains, that in virginity "we have most helps for a *contemplative* life, as in [marriage] we are better furnished for an *Active serviceable* life."[12] Jeremy Taylor, similarly, writes in 1650 that "Virgins must be retired and unpublick."[13] But Marvell's figuration of the more "contemplative" state of virginity loosens the conventional opposition of private virginity and public marriage to introduce us to a new mode of virtuous agency that may be distinct from the active life but that can nonetheless effect a politically wide-scale redemption here in the temporal world. The simple condition of bodily purity promises for Marvell a means of participating in the process of terrestrial reformation. In order better to understand the implications of the complex dialectic structuring the conflict between natural virginity and supernatural marriage, we must examine in all its peculiarity Marvell's rhetorical construction of the distinct types of power Mary exercises over the garden-world of Nunappleton. In representing the contradictory powers Mary derives from her antithetical status as both virgin and bride-to-be, Marvell, I hope to demonstrate, draws on the emotional valences of two distinct political philosophies. Like Harvey, who could justify his conflicting accounts of the circulation only by appealing to the figures of conflicting political systems, Marvell represents the differences between virginity and marriage, and between natural and divine modes of historical agency, as competing modes of political organization. An analysis of Marvell's figural homologies should unfold some of the ideological implications of the idealization of virginity in the years of the English Revolution.

Let us turn to Marvell's rhetorical delineation of Mary's "virginal" effect on the natural development of the landscape. Like a figure in a Spenserian pageant, the innocent Mary enjoys in her progress a ritualized, ceremonial

---

12. Richard Baxter, *Christian Directory* (London, 1673), p. 476.
13. Jeremy Taylor, *Holy Living* [1650], ed. P. G. Stanwood (Oxford: Clarendon, 1989), p. 80.

relation to the natural world. The narrator, as we have seen, calls on the "Fields, Springs, Bushes, Flow'rs" to take advantage of this medial time before Fate's "worthy" but regrettable translation of Mary to the other-worldly state of marriage. This ceremony of natural growth does not, as we might expect, affirm a static or even cyclical view of natural process, that ultimately ordered variety in nature which confirmed for Spenser that all "is eterne in mutabilitie."[14] More like the unidirectional motion of natural process in the prophecies of Winstanley, the elements of the Nunappleton topography are urged to follow a predetermined trajectory toward a "green Age" of horticultural excellence (496), a gradual ontological ascent by which they will surpass all other "*Woods, Streams, Gardens, Meads.*" Marvell, I would argue, follows Winstanley in supplying this image of rectilinear horticultural fecundity with an indisputably political resonance. He isolates in a single symbolic system the private ideal of virginity and the public ideal of the natural reformation of human society.

In articulating the bond between Mary and the burgeoning natural world, Marvell forges a symbolism that is not, however, immediately expressive of political relations in his own troubled time. Marvell's exordium to nature evokes not the Cromwellian Commonwealth but, curiously, the medieval institution of feudalism: "Employ the means you have by Her, / And in your kind your selves preferr" (749–50). The narrator here exhorts the natural elements to behave toward the virginal Mary much as medieval vassals might toward their feudal lord. The features of the landscape receive from their feudal benefactress the freedom and "means" by which they could fulfill the potential of their "kind," their natural position within the ordered hierarchy of the teleologically oriented universe. In describing that period before her emergence on the scene as a future participant in the sacrament of marriage, Marvell figures the process of natural growth in the language of feudal pageantry:

> 'Tis *She* that to these Gardens gave
> That wondrous Beauty which they have;
> *She* streightness on the Woods bestows;
> To *Her* the Meadow sweetness owes;
> Nothing could make the River be
> So Chrystal-pure but only *She*;
> *She* yet more Pure, Sweet, Streight, and Fair,
> Then Gardens, Woods, Meads, Rivers are.

14. *The Faerie Queene*, 3.6.47, in *Spenser: Poetical Works*, ed. J. C. Smith and E. de Selincourt (London: Oxford University Press, 1912), p. 176. Subsequent citations are taken from this edition and cited by book, canto, and stanza number in the text.

> Therefore what first *She* on them spent,
> They gratefully again present.
> The Meadow Carpets where to tread;
> The Garden Flow'rs to Crown *Her* Head.
>   (689–700)

Mary bestows on her topographical subjects the feudal *beneficium* of virginal "streightness," and in emitting their "sweetness" the grateful flowers respond to this magnanimity with the loving and voluntary tribute proper to their kind.

Legal historians of the seventeenth century were just beginning to arrive at a historical understanding of feudalism as an outmoded social institution.[15] Although pledges of vassalage in medieval England may have resulted from threats of economic or even physical force, the feudal bond could nevertheless be articulated in theory as a significant guarantor of virtue and freedom; the feudal distribution of power could be justified in Marvell's time as a "system of mutual responsibilities, a closely interlocking system of classes, each with its unique function, for all of whose good the ruler was responsible."[16] Relying on what is undoubtedly a narrow aspect of medieval political theory, Marvell represents the process of natural growth on the estate as just this kind of "interlocking system," attributing to the virginal Mary the powers of the feudal lord who, in return for fealty and homage, guarantees the security and consensual freedom of his beloved vassals. We can isolate an important reason for which Marvell evokes the ethos of feudal ritual: he wants, I think, to establish the nature of the authority by which Mary reforms her green world. The noble nature empowering Mary does not derive from property or an aristocratic bloodline; her lordly relation to the gardens, woods, meads, and rivers is based quite simply on her own bodily purity. By invoking a feudal ideology whereby the social order is maintained not by a de facto control of military force but by the emanation of a lord's inherent virtue which his vassals freely repay, Marvell establishes the status of virginity as an impalpable but nonetheless natural power. Representing a strength detached from either human violence or the capricious power of divine grace, virginity is a force contingent solely on the innocent condition of the human body.

We have just observed the genial covenant that unites Mary and the gar-

15. J. G. A. Pocock charts the "discovery of feudalism" in *The Ancient Constitution and the Feudal Law: A Study of English Historical Thought in the Seventeenth Century*, 2d ed. (Cambridge: Cambridge University Press, 1987), pp. 70–123.

16. Stevie Davies, *Images of Kingship in "Paradise Lost": Milton's Politics and Christian Liberty* (Columbia: University of Missouri Press, 1983), p. 130.

den in topographical concord and that seems to guarantee the natural process of the garden's reformation. But in his description of the supernatural
power Mary derives from her prospects in marriage, Marvell depicts a violent, apocalyptic vitrification of the natural world that clearly distinguishes
itself from the ritual of covenantal integration that signifies natural growth.
The narrator's initial response to Mary emphasizes her capacity to stun nature with an otherworldly power into rapt attention and servile adoration:

> The gellying Stream compacts below,
> If it might fix her shadow so;
> The stupid Fishes hang, as plain
> As *Flies* in *Chrystal* overt'ane,
> And Men the silent *Scene* assist,
> Charm'd with the *Saphir-winged Mist*.
> (675–80)

For the speaker, at this point in his narration, Mary appears as an object of
apotheosis; she is invested with so much divinity that she may be less identifiable as the thirteen-year-old Mary Fairfax than as a schoolgirl antitype of
Queen Elizabeth, or more dramatically, of the Christian Messiah. Critics
have noted in passing the odd way in which Mary seems to emerge on the
scene much as the triumphant, herculean Son in Milton's "Nativity Ode."
Like the "Shepherds on the Lawn" in Milton's ode who are "in blisfull
rapture took" (84, 98), or like the stars that "with deep amaze, / Stand fixt
in stedfast gaze" (69–70), the rustics on the Fairfax estate "the silent *Scene*
assist, / Charm'd with the *Saphir-winged Mist*."[17] Much as Milton's incarnate
Son silences the "voice" and "hideous hum" of the pagan oracles (173–74),
Mary mesmerizes those around her with her awe-inspiring divinity, be-
numbing the natural world with a "horror calm and dumb" (671).

Insofar as Mary's influence derives from the sacramental action of matrimony, the narrator seems to imagine her an object of idolatrous worship. To
Mary-as-virgin, nature gratefully pays a feudal tribute, while to Mary-as-
future-bride, nature is stunned into the abject servitude more properly reserved for an absolutist Stuart monarch. In a description that contrasts
significantly with the social reciprocity intrinsic to Mary's feudal lordship, the
account of Mary's awe-inspiring triumph conjures the century's earlier Neo-
platonic celebrations of the Caroline monarchy. In labeling Mary the *"Law*

17. The relation of Marvell's depiction of Mary to Milton's "Nativity Ode" is discussed briefly
in Allen, *Image and Meaning*, pp. 220–21; Judith Scherer Herz, "Milton and Marvell: The
Poet as Fit Reader," *MLQ* 39 (1978): 229–63; and Stocker, *Apocalyptic Marvell*, pp. 142–
46.

/ Of all her *Sex*, her *Ages Aw"* (655–56), Marvell adopts the courtly justification of absolutism employed by Thomas Carew, who had praised in similar
terms the beauty of the Caroline queen, Henrietta Maria, "the Aw / of
[whose] chaste beames, doest give the Law" to the natural world.[18] Mary's
triumphant procession as a bride-to-be suggests at once a divine cosmos
controlled by an arbitrary God and a political state governed by a human
sovereign no less arbitrary. In order to reconstruct the ideological implications of Marvell's complex figural distinctions, we might avail ourselves of a
recent comparison of the ceremonies specific to the Renaissance institution
of the absolutist monarchy and the medieval institution of feudalism; in terms
that usefully describe the difference between the two types of authority Mary
exercises on the estate, Jacques le Goff writes: "There are two systems, one
royal, the other 'familial' and aristocratic; two symbolisms, one of transmission of cosmic, supernatural power, the other of familial integration."[19] The
stultifying power of Mary's enchanting presence, however breathtaking its
figuration in Marvell's highly "metaphysical" stanzas, suggests more the tyrannical, supernatural power of absolute monarchy than the aristocratic freedom associated, at least in theory, with the consensual obligation of liberality
and fealty integral to the feudal household.

Although *Upon Appleton House* characterizes virginity and marriage as if
they were alternative political states, Marvell's poem does not, I think, war-
rant a reading as a symptomatic reflection of that historical phenomenon we
have come to see as the two-century-long transition from feudalism to capitalism. Nor need we imagine that Marvell, insofar as he idealizes Mary's
virginity, is laboring under a delusion that England could actually flourish by
a timely reversion to its feudal past. Although we may reject an easy, unmediated isomorphism between Marvell's figural world and a seventeenth-
century material base, the feudal representation of virginity marks Marvell's
participation in one of the important discursive events of the English Revolution. It was the difficult struggle to authorize a discourse of the liberal
organization of the polity that motivated the passionate historical interest in
feudalism, a mode of political organization that provided contemporary political philosophers with a contractual, consensual model of government that
was founded on the putative freedom of both governor and governed. Stevie
Davies has demonstrated persuasively that Milton, in *Paradise Lost*, employs
the newly recovered image of feudal kingship to describe the ideal govern-

---

18. Thomas Carew, "To the Queen," in *Minor Poets of the Seventeenth Century*, ed. R. G.
Howarth (London: Dent, 1953), p. 138.

19. Jacques le Goff, *Time, Work, and Culture in the Middle Ages*, trans. Arthur Goldhammer
(Chicago: University of Chicago Press, 1980), p. 271.

ment structuring both heaven and the unfallen Eden. Much as the meadow in *Upon Appleton House* is seen "gratefully" to present Mary with "Carpets where to tread," the animals in Milton's Eden pay Adam "fealty" (8.344), acknowledging him their lord not for his mystified possession of a legalistic divine right but for his role as the covenantal feudal sovereign, primus inter pares. "Within the unlikely structure of feudal lordship," explains Davies, Milton is "able to demonstrate the deepest meanings of human liberty and equality."[20] It is precisely this identification of the feudal lord as the first among equals, I would suggest, that provided an important conceptual model for the leader of the nascent republican Commonwealth: as Marvell writes of Cromwell in "The First Anniversary," although "Abroad a King he seems," he is "At Home a Subject on the equall Floor" (389–90). Marvell's Cromwell fashions his role of Protector as the feudal lord par excellence; like the Harveian heart in 1649, Cromwell may seem a king, but this tentative monarchy is absorbed into a larger structure of egalitarian organization.

Despite, however, the surprising implications of republicanism borne by the poem's invocation of the aristocratic ceremony of vassalage, we can see as well a local and perhaps more pressing concern behind Marvell's feudal depiction of Mary's virginal powers. Marvell establishes the proper social milieu for human virtue and political reformation by locating his figural ideal—the magical process of natural reformation—in the institution central to the feudal mode of production, the household. In 1655, Richard Baxter had advised his fellow ministers, "You are like to see no general reformation till you procure family reformation."[21] By invoking the superannuated feudal world in describing Mary's relation to nature, Marvell conjures the Puritan movement toward the liberalization of the polity informing Baxter's dictum. The feudal covenant binding the virginal Maria to a gracious natural world functions as a highly mediated figure for the newly idealized political organization in which all social relationships are based on the theoretically noncoercive and noncentralized model of a nation of feudal households.

We have seen how the earlier, sacramental description of Mary-as-future-bride reinforces a traditional sense of individual and polity at odds with the radical humanism at the heart of the Marvellian poetic. The apocalyptic figuration of Mary's mesmerizing power points theologically to man's helpless resignation to an awe-inspiring, arbitrary divinity that may or may not reward man's sacrificial actions; and it points politically to man's unthinking subjec-

20. Davies, *Images of Kingship*, p. 163.

21. Richard Baxter, *The Reformed Pastor* (London, 1655); quoted in Christopher Hill, *Society and Puritanism in Pre-revolutionary England* (London: Continuum, 1964), p. 445.

tion to the awe-inspiring, arbitrary monarch. The description, however, of Mary as virginal student, immersed in feudal reciprocity with the natural world, affirms a uniquely liberal vision of both God and polity: it suggests the believer's capacity to participate in the redemptive forces of a laboring creation and the citizen's right to participate in a polity established by freely willed consent. Through this complex, dialectical representation of Mary's marriage, Marvell arrives at an extraordinary alternative to the culture's dominant view of political change as the result of either militant human action or transcendent divine will. He attempts to locate in the natural world a vision of revolution whose causal mainspring is not the blatantly manipulative acts endemic to the polity but the benign, virtuous activity nestled comfortably within the household. The feudal organization of the green world represents an attempt to establish a *tertium quid* between the troubling poles of the *vita activa* and the *vita contemplativa*. Marvell creates a world in his pastoral works in which retirement represents something quite different from a gross dereliction of public responsibility; rather than a temporary refuge from the important activity in the society at large, virginal retirement represents, at least temporarily, the basis for a revolutionary conception of both human agency and political organization.

A Eunuch for the Kingdom of Heaven

Thus far we have considered the implications of Marvellian virginity by pursuing the outlines of the political organization it seems to evoke. We now direct our focus to Marvell's interest in a set of concerns specific to virginity as a state of sexual abstinence. In order better to understand the peculiar figural logic by which Marvell employs the image of Mary Fairfax's virginity to assert the new power of passive private virtue, I propose that we examine the possibility of Marvell's reworking of an analogous nexus of concerns from Pauline theology.

The perennial conflict between free will and necessity has provided many Western cultures with an important conceptual vocabulary for the construction of discursive systems that we can identify as politically motivated philosophies of agency. It is an avatar of the dialectic of free will and necessity that supplies St. Paul with an important philosophical focus for his millenarian anticipation of "the redemption of our bodies." I think we can see how the representation of the Fairfaxes' commitment to make "*Destiny* their *Choice*" indicates Marvell's explicit engagement of just this conflict in historiographical thought. Critics of Marvell speak often of Fairfax's commitment to a "providential destiny," but this phrase would have been dismissed as oxymoronic by Marvell, as it would have by St. Paul.[22] We can understand the bifurcation of providence and destiny in Marvell if we examine Fairfax's concession to his daughter's marital destiny against the background of an important Pauline text on man's preparation for redemption. Behind Lord Fairfax's commitment to marry off his virgin daughter looms a scriptural injunction against just such a commitment—Paul's pronouncement that marriage is a form of social destiny that it is the duty of any independent Christian, seeking redemption, to avert. As if in muted counterpoint, we can hear behind the marriage stanza in *Upon Appleton House* Paul's advice to the father who feels constrained by social pressures to give his virgin daughter away in marriage:

> If any man think that he behaveth himself uncomely toward his virgin, if she pass the flower of her age, and need so require, let him do what he will, he sinneth not: let them marry. Nevertheless he that standeth steadfast in his heart, having no necessity, but hath power over his own will, and hath so decreed in his heart that he will keep his virgin, doeth well. So then he that giveth her in marriage doeth well; but he that giveth her not in marriage doeth better. (1 Cor. 7:36–38)

Paul acknowledges that some concessions to the societal pressure to marry must, of course, be countenanced. Nevertheless, he encourages the father to avoid a typically Roman, Stoic resignation to a social and historical "necessity" that may seem on the surface to lie beyond the father's control. For a father to "keep his virgin" within his own household even after she has passed "the flower of her age" can in some way constitute not a choice for, but the proper defiance of, the destiny imposed by the public world of social and political obligation. The Christian who resists giving his daughter in marriage exercises, for Paul, "a power over his own will," challenging the accepted codes of patriarchal behavior that work all too efficiently to determine the course of events.

Elaine Pagels has neatly summarized the way in which the early Christian ethic of private renunciation first established itself in opposition to the dominant Stoic discourse of destiny and public obligation: "The Christians repudiated what Marcus Aurelius regarded as the highest virtue, for . . . Marcus saw his religious destiny given in his familial, social, and political situation, and in the duties his imperial role placed upon him."[23] The practice of sexual abstinence became attractive to early Christians because it was seen

---

22. See, for example, Stocker, *Apocalyptic Marvell*, p. 66; and Grossman, "Authoring the Boundary," pp. 199, 201.

23. Elaine Pagels, *Adam, Eve, and the Serpent* (New York: Random House, 1988), p. 84.

to free them from a connection to the public world, from those socially
determined forces of "fate" and "necessity." Paul, in another incitement to
virginity, wrote that "she that is married careth for the things of the world,
how she may please her husband," whereas the "unmarried woman careth
for the things of the Lord" (1 Cor. 7:34). For many in the late classical
period, the unyielding sexual abstinence of unmarried adult women represented a scandalous repudiation of the things of this world, an egregious
refusal to participate in the larger arena of public and social concern.[24] Exploiting and elaborating on the social implications of the patriarchal fear of
sustained virginity, the Pauline epistles suggested that sexual renunciation
was not just a negative act of public defiance but a positive and ultimately
efficacious private virtue. In those moments of the figural idealization of
virginity in *Upon Appleton House*, Marvell, I suggest, adopts the Pauline
association of sexual abstinence with the possibility of the virgin's freedom
from the world of marital, and thereby political, destiny; her freedom, in
other words, to ally her passive virtue with the laboring forces of an organic
revolution.

We have examined the symbolic domain Marvell employs to represent the
assertions of self in the realms of both sexual and civil behavior: both sex
and politics are subsumed into a more general category of action, the imposition of human will onto a pacific and self-sufficient realm of nature.
Although Marvell's images of sexual activity body forth man's action in the
polis, a recognizable social referent for sexual abstinence is not so forthcom-
ing. By examining Marvell's juxtaposition of Mary's virginal repudiation of
erotic combat with Fairfax's retreat from his military obligation, we can explore how virginity in Marvell's verse widens to encompass some important
modes of politically resonant behavior. We have observed the care with
which Marvell distinguishes between the mysterious "universal good" that
may arise from matrimonial sacrifice and the more local effects of natural
efflorescence contingent on virginity. The distinction he makes here suggests
inevitably the two types of reformation available to Fairfax in 1650:

> And yet there walks one on the Sod
> Who, had it pleased him and *God*,

---

24. See Peter Brown, *The Body and Society: Men, Women, and Sexual Renunciation in Early
Christianity* (New York: Columbia University Press, 1988), pp. 80–81. In his discussion of the
subversive social threat that sexually abstinent women were seen to pose in the late classical
period, Brown relays the patristic conviction that the mystical powers boldly assumed by self-
proclaimed virgins were but impious assertions of "human initiative," since virginity was a condition not of grace but of nature.

> Might once have made our Gardens spring
> Fresh as his own and flourishing.
>
> (345–48)

In *Upon Appleton House*, historical progress is imaged as horticultural fecundity, and the question with which the poem so often struggles is whether
it should be *"our* Gardens" or one's *"own"* that one chooses to make spring
and flourish; whether it is for a "universal good" or for the good of one's
private household that one fashions one's behavior.

When Fairfax retreated from the political action that might have cultivated
the garden of England, he made his own garden spring fresh and flourishing.
This epochal choice between the spheres of action and retirement is figured
as the gardener's discrimination between two plants: "For he did, with his
utmost Skill, / *Ambition* weed, but *Conscience* till" (353–54). It is this image
of the superior plant of *"Conscience"* through which Marvell identifies the
moral principle indigenous to the garden-state of retirement and draws together the retired Fairfax and the virginal Mary as analogous historical
agents. Like Mary, whose virginal purity derives from her origins in her
"Domestick Heaven," Fairfax's plant of conscience has been nursed and
nourished in the Christian heaven:

> *Conscience*, that Heaven-nursed Plant,
> Which most our Earthly Gardens want.
> A prickling leaf it bears, and such
> As that which shrinks at ev'ry touch;
> But Flowrs eternal, and divine,
> That in the Crowns of Saints do shine.
>
> (353–60)

We can best pursue the connection Marvell forges between Fairfax's
"Conscience" and the virginity of his daughter once we acknowledge that
this plant is itself a scion of an earlier literary plant. George deF. Lord
has noted the indebtedness of Marvell's vegetative *"Conscience"* to the
magical "Haemony" of Milton's *Comus*, and an examination of the Miltonic
original may help us account for the remarkable idiosyncrasy with which
Marvell represents his chaste ethical ideal.[25] The Attendant Spirit in
Milton's masque presents the Lady's two brothers with a magical herb that

---

25. George deF. Lord, ed., *Complete Poetry*, by Andrew Marvell (London: Dent, 1984), p.
74n. Marvell has also, I think, transplanted the Fairfacian "conscience" from the third book of
The Faerie Queene, the Book of Chastity, in which the narrator, praising the virginal Belphoebe,
extols that "faire flowre / Of Chastity and vertue virginall" that will "crowne your heades with

can shield them from the temptations of Comus; of the virtuous plants and healing herbs shown him by "a certain Shepherd Lad," Thyrsis remarks:

> Amongst the rest a small unsightly root,
> But of divine effect, he cull'd me out;
> The leaf was darkish, and had prickles on it,
> But in another Country, as he said,
> Bore a bright golden flow'r, but not in this soil.
>
> (629–33)

Like Marvell's "Heaven-nursed" conscience, Milton's virtuous plant flourishes best in other lands, as the fallen soil of this realm does not encourage its bloom. Surely the most striking aspect of Milton's characterization of this "unsightly root" culled by the Shepherd Lad is its strong suggestion of masculine sexual potency: "Haemony," after all, might have an etymological tie to *haimonios*, blood-red. Dark and prickled, this phallic plant exercises its virtue by means of *virtù*, embodying the rigid law of the father that requires ultimately not virginity but the state of "married chastity."[26]

Although it has been argued that "Marvell does not seem at all interested in the programmatic center of *Comus*,"[27] *Upon Appleton House*, as we have seen, clearly shares with Milton's masque a considerable interest in the mor-

---

heauenly coronall" (3.5.53). Much as Marvell's *"Conscience"* is a "Heaven-nursed Plant, / Which most our Earthly Gardens want," so in *The Faerie Queene*:

> Eternall God in his almighty powre,
> To make ensample of his heauenly grace,
> In Paradize whilome did plant this flowre,
> Whence he it fetcht out of her natiue place,
> And did in stocke of earthly flesh enrace.
>
> (3.5.52)

26. Richard Halpern, "Puritanism and Maenadism in *A Mask*," in *Rewriting the Renaissance: The Discourses of Sexual Difference in Early Modern Europe*, ed. Margaret W. Ferguson, Maureen Quilligan, and Nancy J. Vickers (Chicago: University of Chicago Press, 1986), p. 104. I would argue, in addition, that both Milton and Marvell rely on the venerable literary associations borne by the *agnus castus*, or "chastity plant," an actual species with "digitate leaves and spikes of purplish blue" which has been associated with virginity from Plato through the Church Fathers (Herbert Musurillo, ed., *The Symposium: A Treatise on Chastity*, by Methodius [London: Longmans, 1958], p. 186n). Of the *agnus castus*, around which the virgins of Methodius's *Treatise on Chastity* convene, the character Marcella claims: "It was indeed a most extraordinary disposition that the plant of virginity was sent down to mankind from heaven" (pp. 43–44).

27. Herz, "Milton and Marvell," p. 261.

al force Milton's Lady calls the "Sun-clad power of Chastity" (782). But in order to understand why the conscience Fairfax cultivates in *Upon Appleton House* should bear this association with virginity and chastity, we must recognize how Marvell's plant subjects Milton's haemony to a crucial revision. Marvell, indeed, seems to reproduce in his description of Fairfax's conscience Milton's figurative sexualization of this "prickled" plant of virtue. But the younger poet has also quite explicitly drained Milton's image of its sexual potency and assertiveness: Marvell's *"Conscience"* may, like Milton's haemony, bear a "prickling leaf," but Marvell's plant is one that "shrinks at ev'ry touch." With this suggestion of a detumescent masculinity that shies instinctively away from active engagement, Marvell is able to represent as his moral ideal an appropriately masculine version of the Lady's commitment not to married chastity but to the "sage / And serious doctrine of virginity" (786–87). Like his daughter's, Fairfax's *"Discipline Severe"* combines conscience and continence into a powerful moral unity.

That Marvell has attempted to charge Fairfax's conscience with the traditionally "feminine" ethos of virginity is suggested further, I propose, by an analogous figure in an early poem attributed to Marvell, "An Elegy upon the Death of my Lord *Francis Villiers.*" At some point in nearly all his public panegyrics, Marvell situates the heroic personage most unheroically in the private sphere, and the elegy to Villiers is no exception. Before stepping into the battle where he would soon lose his life, the Villiers of Marvell's poem anticipates a regenerative sexual encounter with his wife, the "matchlesse *Chlora* whose pure fires . . . only could his passions charme" (69–70). Villiers's wife, however, in an exquisitely Marvellian fashion, resists these advances: neglecting her husband's desires "for honours tyrannous respect," Chlora, "like the Modest Plant at every touch / Shrunk in her leaves and feard it was too much" (79–82). The plant of conscience so carefully tended by Fairfax, I submit, should be scrutinized as a politicized offshoot of just this shrinking "Modest Plant" of retiring, and implicitly feminine, virginity.[28]

---

28. The undeniably quirky nature of Marvell's sexual imagery has led many of his strongest critics to pursue its "Cause unknown" in the author's psyche. Attempting to account for what seems the poet's excessive interest in young girls, William Kerrigan, "Marvell and Nymphets," *Greyfriar* 27 (1986): 3–21, examines Marvell's lyrics in light of the twentieth-century recognition of the horrors of child molestation, leveling the charge of an almost Nabokovian pedophilia. On the other hand, Leah Marcus, in *Childhood and Cultural Despair: A Theme and Variations in Seventeenth-Century English Literature* (Pittsburgh: University of Pittsburgh Press, 1978), p. 228, has ventured that Marvell's youthful interests stem from an actual physical condition of

Through the allusion to *Comus* in the description of Fairfax's heroic conscience, Marvell announces his appropriation of the dialectic of chastity and
virginity so central to Milton's masque. Critics of *Comus* have mused for
some time on the Lady's shift in emphasis from the "chastity" of which her
brothers speak to the actual bodily condition of her virginity.[29] It is likewise
not the more traditional Protestant ideal of "married chastity" that comprises the Marvellian ideal but the far more conservative, even Catholic,
identification of chastity with the physical state of virginity. But despite this
particular instance of indebtedness to the sensibility of Milton's Lady, Marvell's particular figuration bespeaks an entirely new characterization of virginity as both a physical state and a moral ideal. In Milton's *Comus*, the
natural pleasures to be found in the sensual world present a genuine temptation, and the practice of chastity is figured, therefore, as vigilant and active self-restraint, a judicious exercise of the will that excludes, rejects, and
denies the welling up of natural passions. The Lady prides herself on her
strict adherence to Nature's "sober laws, / And holy dictate of spare temperance" (765–66); and with this conception of virginity as the result of
judicial prohibition, Milton follows fairly closely the major Western figurations of sexual abstinence. Marvell, however, loosens virginity from its classic identification with a negatively induced program of sexual renunciation.
Virginity is something less to be struggled for than it is to be embraced,
caressed, and even, if we consider a contemporary's praise in 1650 of "this
delicious perfume of virginity," inhaled.[30] In a significant departure from the
standard paradigm of the negation of libidinal energies, Marvell lends virginity the shape of the "apparatus of sexuality" described by Foucault, the
social process whereby a seemingly repressive phenomenon such as sexual

---

impotence. However shaky the historical grounds for Marcus's supposition, her biographical
reading at least provides the more compelling insight into the poems: "Marvell portrays adult
sexuality much as little children are likely to see it—as brutal violence, a form of self-injury or
even death.... He shies away from feeling physical desire along with his esteem for the love
object and seeks out the love of little girls with whom adult sexuality is out of the question."
See also William Empson, *Using Biography* (London: Chatto and Windus, 1984), pp. 14–16,
78–81, 86–87. In this notorious and elaborate hypothesis of the poet's sexual constitution, Empson propounds his theory, quite cleverly at times, of Marvell's occasional homosexuality ("I think
he fell in love with the Mower").

29. The Lady's and, by extension, the young Milton's attachment to an ideal of sexual abstinence has been questioned by John Leonard, "Saying 'No' to Freud: Milton's *A Mask* and
Sexual Assault," *Milton Quarterly* 25 (1991): 129–40, and properly reasserted by William Kerrigan, "The Politically Correct *Comus*: A Reply to John Leonard," *Milton Quarterly* 27 (1993):
149–54.

30. Nicholas Caussin, *The Holy Court in Five Tomes*, trans. T[homas]. H[awkins]. (London,
1650), p. 106.

abstinence may indicate not "the rejection of sex, but a positive economy of the body and of pleasure."[31]

The Marvellian virgin, in fact, enjoys an intensification of bodily pleasure, in an untempered surfeit of the innocent pleasures of nature that are best enjoyed alone. Whereas the promise of a shared sexual experience—or even the less physical activity of chaste erotic courtship—bodes only peril, a true indulgence in the pleasures of nature is granted first and foremost to the virgin. Rejecting the characterization of virginity as the sober adherence to the puritanical canons of self-control, Marvell writes in praise of a bodily condition that borders on an unfettered libertine sensuality. Whereas Milton's *Mask* traces bodily pleasure to the material evils embodied by Comus, Marvell seems to assert that it is the virgin who alone can appreciate the fruits of what Comus describes as Nature's "full and unwithdrawing hand" (720–24). Marvell locates virginity in that ideal pastoral realm of isolate sensuality envisioned in "The Garden," the retreat that may allow for a passive and solipsistic erotic languishing but that eschews the rough strife of sexual assertiveness—or assertiveness, for that matter, of any kind.

Walter Charleton, in his youthful enthusiasm at the Vitalist Moment, could imagine a means of human agency that was "not obliged to the dull condition of an immediate corporeal contact."[32] This scientific, vitalist fantasy of non-contactual agency was extended by Sir Thomas Browne to a fantasy of non-contactual reproduction: "I could be content," writes Browne in *Religio Medici*, "that we might procreate like trees, without conjunction, or that there were any way to perpetuate the world without this triviall and vulgar way of coition."[33] John Evelyn, too, expresses an envy of the way in which trees can "generate their like . . . without violation of virginity."[34] These wistful fantasies of a vegetable love that avoids the vulgarity of the physical impact necessary for sexual coupling seems to find representation in Marvell as a palpable, realized alternative. The exquisitely narcissistic speaker of *Upon Appleton House*, "languishing with ease" and tossing "On Pallets swoln of Velvet Moss," incamps his mind in a protective, solitary grove in which all nonvegetative seductions are rendered limp and ineffective: "Where Beauty, aiming at the Heart, / Bends in some Tree its useless Dart" (603–4). During the Restoration, some of Marvell's more rancorous political en-

31. Michel Foucault, *Power/Knowledge*, ed. Colin Gordon (Brighton: Harvester, 1980), p. 190.

32. See van Helmont, *Ternary of Paradoxes*, sig. D4ᵛ–E1ʳ; quoted in Gelbart, "Walter Charleton," p. 155.

33. Thomas Browne, *The Major Works*, ed. C. A. Patrides (Harmondsworth: Penguin, 1977), p. 148.

34. Evelyn, *Sylva*, 2:263.

emies chose to ornament their ideological objections to the satirist with attacks on his sexual identity. The line from one anonymous poem, labeling
Marvell the "Gelding, and Milton the Stallion," whatever its value as biographical conjecture, sums up quite neatly the poetic stances toward sexuality
assumed by the two writers.[35] Whereas sex in Milton is a temptation that
the impassioned subject must heroically resist, a weakness he must submit
to the "holy dictate of spare temperance," sex for Marvell is but a cruel twist
of fate—the necessity Browne recognized as the need for procreation—to
which the individual must be forced to concede. In considering sexuality in
Marvell's poetry, we cannot with much accuracy speak of the "practice" or
"exercise" of virginity, as the natural, passive state of bodily purity requires
no purposive thought. The Marvellian hero has no call to rise to the heroic
efforts of self-restraint required of Milton's heroes; Marvell's moral and political ideal, rather, is best characterized by the touching image of a bashful
flaccidity, the hero as the trembling, self-enclosed, self-delighting drop of
dew.

We can at this point begin to trace a consistent pattern behind Marvell's
allusions to Milton's *Poems* of 1645, a revisionary pattern that recapitulates
the process of demasculinization we have observed in Marvell's revision of
Milton's haemony and that forms the basis of Marvell's unique poetics of
passive agency. Critics have noted a correspondence between Milton's image
of the sun in the "Nativity Ode" and Marvell's characterization of the sunlike
poet-tutor near the end of *Upon Appleton House*. The sun in Milton's poem
is a decidedly masculine rake, the "lusty paramour" of "Nature" (36), who
finds himself, at the Incarnation of Christ, in a retreat to his solar boudoir:
retiring to a bed "Curtain'd with cloudy red," the sun "Pillows his chin upon
an Orient wave" (230–31). Transmuting this Miltonic figure for the sun's
evening retirement upon the arrival of the Son of God, Marvell fashions an
image expressing the speaker's own relation to the apocalyptic glories of the
approaching Mary:

> The *Sun* himself, of *Her* aware,
> Seems to descend with greater Care;
> And lest *She* see him go to Bed;
> In blushing Clouds conceales his Head.
> (661–64)

As Marvell transforms Milton's "lusty paramour" into a modest and, of
course, virginal figure for the poem's own shamefaced poet-tutor, the cloudy

35. Quoted in Lynn Enterline, "The Mirror and the Snake: The Case of Marvell's 'Unfortunate Lover,'" *Critical Quarterly* 29, no. 4 (1987): 109n.

red curtains of the sun's rakish bedroom in the "Nativity Ode" become the "blushing Clouds" that express a quintessentially Marvellian modesty before the vision of Mary. The herculean, implicitly masculine strength with which Milton endows so many of his heroic figures, from the Son in the "Nativity Ode" to Samson, undergoes a Marvellian metamorphosis into the more "feminine" quality of passive virtue. And these tender traits of modesty and gentle introspection that comprise the ethical ideal of *Upon Appleton House* reveal their distance from the unlikely militaristic might of Milton's "dredded Infant" who "Can in his swadling bands controul the damned crew" (220–28).

This local instance of Marvell's feminization of a Miltonic image signals the broader program of Marvell's revision of the Miltonic poetics of political agency. The studied inversion of Milton's masculine images denotes a more significant rejection of the masculinist heroic ideal frequently expressed in Milton's early prose tracts. In *The Reason of Church-Government*, for example, Milton voices a commitment to the active desire "to push [reformation] forward with all possible diligence and speed."[36] In rejecting this revolutionary faith in action celebrated by "Milton the Stallion," Marvell the "Gelding" locates his own ethical ideal in that retiring figure named in "Musicks Empire" as the "gentler Conqueror," Fairfax himself. Conflating the life of retirement with the state of sexual renunciation, Marvell embraces an abstinence that emerges most vividly as the actual state of sexual impotence; the virtuous passivity of the contemplative life is characterized as a male body, potent, perhaps, in isolation, but which "shrinks at ev'ry touch." Fairfax left the realm of public action to nurse his conscience in the private sphere, a retreat into political impotence that stands in direct contrast—in both the political and the sexual sense—to the ambitious drive that would require Cromwell, as Marvell observes in the "Horatian Ode," to "keep [his] sword erect" (116).

We need not diverge widely from the traditional critical assumption that Marvell's insistent feminization of the retired Lord General bespeaks, at one level, an undeniably pointed critique of his patron's political efforts. It is not without reason that Rosalie Colie has mused on Marvell's depictions of Fairfax "in terms of the extraordinary women in his family": "An odd way, one might think, to present a warrior-hero, so domestically and cosily among his women folk."[37] While acknowledging that Marvell seems to consign Fairfax

36. Milton, *Complete Prose Works*, 1:800.

37. Rosalie Colie, *"My Ecchoing Song": Andrew Marvell's Poetry of Criticism* (Princeton: Princeton University Press, 1970), pp. 220–21. Marvell was not alone in feminizing the Lord General. It is curious to conjecture whether it was Fairfax's genuinely modest character or an early reading of Marvell's poetic depictions of him that inspired the man who became Mary's husband to characterize Fairfax so explicitly as a feminized hero. In "An Epitaph on THOMAS,

to a fairly ignominious identification with the household realm, we can also chart a coincidental transvaluation of the household sphere into a locus for potential reformation. Marvell transforms a preternatural sexual modesty and an attachment to the home into a political virtue, and we might trace the remarkable path along which he effects this ideological metamorphosis by returning to the political significance given virginity by St. Paul. For the early Christians, the "destiny" and "fate" that were largely determined by social and familial obligations were to be distinguished from the promised end of Christian history. In fact, to avoid one's marital destiny through sexual abstinence was in many ways to court a superior fate in the Christian millennium. Although the sexualist discourses in the early modern period reveal a disproportionate concern with the chastity of young women, Marvell's representation of Fairfax's "*Conscience*" seems to rely on the Pauline association of the approach of the millennium with the sexual abstinence of both genders. For Paul, it is the passive virtue of virginity, rather than any direct political action, that can best prepare even the patriarch for the eschaton: "the time is short: it remaineth, that . . . they that have wives be as though they had none . . . for the fashion of this world passeth away" (1 Cor. 7:29).

As perhaps best evinced by his Latin verses, "Upon an Eunuch: A Poet," Marvell seems to have been drawn to the curious ideal of hypercompensated emasculation, the early Christian idealization of those men "which have made themselves eunuchs for the kingdom of heaven's sake" (Matt. 19:12). Paul's chiliastic theology, though it considered at times the possibility of a redemption of nature, led more often to the anticipation of an apocalypse that would transcend the petty reformations effected in society, a final divine event that would jettison the familiar temporal world into eternity. Later Church Fathers, however, seized on the potential political implications of the Pauline doctrine of virginity, asserting the relation of virginity to the anticipated "redemption of our bodies," the gradual process of change whereby the entire creation is redeemed on this side of eternity. The third-

---

third Lord Fᴀɪʀꜰᴀx" (1671), the Lord General's son-in-law, George Villiers, second duke of Buckingham, praises "Fairfax the valiant" whose "modesty still made him blush": "He had the fierceness of the manliest mind, / And eke the meekness too of womankind" (I). See *A Third Collection of . . . Poems, Satires, Songs, &c. against Popery and Tyranny* (London, 1689); Buckingham's "Epitaph" is reprinted in *Stuart Tracts, 1603–1693*, ed. C. H. Firth (New York: Cooper Square, 1964), pp. 399–402. See Christopher Kendrick, "Milton and Sexuality: A Symptomatic Reading of *Comus*," in *Re-membering Milton: Essays on the Texts and Traditions*, ed. Mary Nyquist and Margaret W. Ferguson (New York: Methuen, 1987), pp. 43–73. Kendrick rightly identifies the imputation of femininity as a corollary of male virginity: "The reduction of chastity to virginity . . . goes far toward making chastity absolutely feminine, and femininity absolute. Not that only a woman can be chaste or feminine; what tends to be implied, instead, is that a chaste man is feminine, or *identifies with* the feminine in so far as he is chaste" (p. 63).

century celibate Methodius, for example, entertains the possibility of a terrestrial, and therefore political, apocalypse when his character Theopatra asserts that "nothing is superior to chastity in its power to restore mankind to Paradise."[38] For Gregory of Nyssa, it is through virginity that it "is possible to return to the original blessedness."[39] Marvell, in attributing to the Fairfaxes' virginity an effusion of magical power on the landscape, suggests precisely this post-Pauline faith in the role of virginity in the reestablishment of the terrestrial paradise.

I have argued elsewhere that the vitalist movement of the 1640s and 1650s spawned a new interest, despite the period's orthodox Puritan affirmation of marriage, in what the Pauline tradition established as the revolutionary potential of virginity.[40] In one of the central texts of the Vitalist Moment, van Helmont's 1648 *Ortus Medicinae*, sexual abstinence becomes a central feature of an entire theology: "the Almighty hath chosen his Gelded Ones, who have Gelded themselves for the Kingdom of God its sake. . . . those only shall *rise again changed*, who shall *rise again glorified* in the Virgin-Body of Regeneration."[41] The hermetic philosopher Cornelius Agrippa, in a text translated in England in 1651, argued for the benefits of virginity for alchemists and seers and claimed, in his treatise on original sin, that God would admit only virgins into heaven.[42] Even the Anglican Jeremy Taylor joined the midcentury revitalization of virginity, challenging the prevailing affirmation of marriage to advise his flock, in 1650, "to expect that little coronet or special reward which God hath prepared (extraordinary and besides the great Crown of all faithful souls) for those *who have not defiled themselves with women, but follow the* Virgin *Lamb forever.*"[43] The focus of the vitalist revival of virginity rested, I believe, on an interest in the specifically physiological qual-

38. Methodius, *Symposium: A Treatise on Chastity*, IV, 2, p. 75.

39. Gregory of Nyssa, *On Virginity*, in *Ascetical Works*, trans. Virginia Callahan (Washington, D.C.: Catholic University of America Press, 1967), p. 59; quoted in R. Howard Bloch, "Chaucer's Maiden's Head: 'The Physician's Tale' and the Poetics of Virginity," *Representations* 28 (1989): 116.

40. John Rogers, "The Enclosure of Virginity: The Poetics of Sexual Abstinence in the English Revolution," in *Enclosure Acts: Sexuality, Property, and Culture in Early Modern England*, ed. Richard Burt and John Michael Archer (Ithaca: Cornell University Press, 1994), pp. 229–50.

41. Van Helmont, *Ortus Medicinae*, translated as *Oriatrike, or Physick Refined* (London, 1662), p. 670.

42. Henry Cornelius Agrippa, *Three Books of Occult Philosophy*, trans. J[ohn]. F[rench]. (London, 1651), pp. 512–18 and 522–26. For a reading of Agrippa's theology of virginity, see William Kerrigan, "The Heretical Milton: From Assumption to Mortalism," *English Literary Renaissance* 5 (1975): 141; and James Grantham Turner, *One Flesh: Paradisal Marriage and Sexual Relations in the Age of Milton* (Oxford: Clarendon, 1987), pp. 150–62.

43. Taylor, *Holy Living*, p. 74. See also the radical sectarian Roger Crab, whose *Dagons-Downfall* (1657) charts the apocalyptic political effects of celibacy and vegetarianism.

ity of the virginal body, what van Helmont calls, in a scientific discussion of Christ's mother, the "material substance of the Virgin."[44] Situating the Pauline idealization of bodily purity within the framework of this new animist materialism, Marvell elaborates on the physical dynamics of a world in which matter and morality are inseparably intermixed. The state of virginity intersects with the prospect of a natural redemption through their shared contact with the realm of the bodily. In a pastoral world in which human and vegetable development can potentially obey the same laws of nature, the maintenance of virginal integrity, a virtue whose physiological makeup exerts a palpable effect on the physical world, can be seen as the first step in an "apocalyptic" process that redeems the fallen natural world and, by extension, the fallen nature of political organization.

We can, I think, safely assert that Marvell subjects to some scrutiny many of the most widespread assumptions about masculine and feminine sexuality and their relation to power and the capacity for change: he subordinates, to adopt the language of "The Garden," "the uncessant labours" of the polis to the "Fair Quiet" and "Innocence" of the private sphere. At the same time, however, there is quite obviously no move to challenge directly the basis of patriarchal authority. The writings of St. Paul perhaps best exemplify a radical subversion of cultural presuppositions about sexuality which nonetheless reinforces the logic of sexual hierarchy. *Upon Appleton House*, in a similar fashion, derives Fairfax's authority as the gentler conqueror from his role as father and householder. Marvell's implicit patriarchalism, however, marks a significant departure from the philosophy of rule we find in the *Patriarcha* of Robert Filmer. Instead of attempting, like Filmer, to justify state power through its derivation from the rule of family authority, Marvell lays stress on the disparity, rather than the correspondence, between the two realms: the poet's purpose is to reconceive a social totality that emanates from the family community. In this universal extension of what he posits as the moral and sexual innocence of the private sphere, a world in which virtue and not strength is the source of individual agency, Marvell has figured an idealized, spiritualized map of what was slowly taking shape among radical intellectuals in the 1650s as the liberal polity.

"Paradice's only Map"

Although Mary's eventual marriage brings with it the hope of "some universal good," Marvell is careful to indicate that the remarkable natural efflo-

---

44. Van Helmont, *Oriatrike*, p. 670.

rescence she effects is contingent on the maintenance of her virginal state. In *Upon Appleton House*, it is the retirement to the state of virginal inno- cence that forms the basis of the historical progress—an Aristotelian fulfill- ment of potential pastoral excellence—envisioned in the poem's penultimate stanzas:

> For you *Thessalian Tempe's Seat*
> Shall now be scorn'd as obsolete;
> *Aranjuez*, as less, disdain'd;
> The *Bel-Retiro* as constrain'd;
> But name not the *Idalian Grove*,
> For 'twas the seat of wanton Love;
> Much less the Dead's *Elysian Fields*,
> Yet nor to them your Beauty yields.
>   (753–60)

As the historical process of pastoral perfection, a movement that contains a chastening supersession of the seat "of wanton Love," is contingent on the condition of virginity, Nunappleton can only supersede these celebrated gar- dens now, while Mary still "leads her studious Hours" in virginal retirement. It is only "now" (754) in the "mean time" of Mary's virginal state, that the narrator can salute the Nunappleton estate, once fallen in war, as fully re- deemed: "*You Heaven's Center, Nature's Lap. / And Paradice's only Map*" (767–68).

Despite, however, this rather startling paean to the powers of virginity, Marvell seems to voice a deeper anxiety concerning the geographical extent of the redemptive potential embodied in Mary's private virtue. Given the fact that her virtue can impel Nunappleton to an apex of reformation, can the cultivation of virginity, or, for that matter, of any private virtue, reform the public world that lies outside the confines of the household? Or is it only an extraordinary action in the polis—the sacrifice of one's private pleasures for martial or marital ends—that can redeem the fallen garden of England? These are the questions about the social consequences of private morality to which Marvell returns throughout his career and to which he gives a variety of answers. It is surely no accident that the solution to these questions yielded in *Upon Appleton House* is embedded in a syntactical ambiguity. After extolling the redemptive process that embraces Fairfax's estate, Marvell has added:

> 'Tis not, what once it was, the *World*;
> But a rude heap together hurl'd;

> All negligently overthrown,
> Gulfes, Deserts, Precipices, Stone.
> Your lesser *World* contains the same.
> But in more decent Order tame.
>   (761–66)

The generally progressive mood of the stanzas immediately preceding this passage works to imply that the "lesser *World*" of Nunappleton has been able to "contain" the greater "*World*" outside, wrapping its beneficent household virtue around the society at large, drawing the chaotic public world into its "more decent Order tame." In this reading Marvell's phrase, "a rude heap together hurl'd," names the world as "once it was," the chaos to which Mary can impart the beauty and order of her own virginal excellence. Through the girl's local practice of virtue and virginal modesty, the chaotic "Gulfes, Deserts, Precipices, Stone" of the world at large have gradually proceeded, in a ceremony of redemption, toward the state of perfection attained by the "*Woods, Streams, Gardens, Meads*" within the lesser world of the Fairfax estate. Marvell, it could be argued, suggests that the floral progress during the "mean while" of Mary's virginity has somehow extended its redemptive scope to the rude world beyond the domestic sphere.

It must be acknowledged, however, that this stanza can also invite a reading quite contrary to this mood, which Harry Berger has properly labeled a "Baconian optimism."[45] By the logic of Marvell's loosely appositional syntax, there lingers the possibility that *rude heap* names not the world "as once it was" but the world as it is now. And if we credit this grammatical alternative, we must alter significantly our understanding of the historical progress Marvell describes. If "rude heap" describes the world in the present, the stanza is sketching a narrative whereby the chaotic forces of the Civil War have reenacted the Fall, negligently overthrowing what was once the orderly, "unfallen" world of the English countryside, leaving unspoiled, as if for a quaint reminder, the "lesser" world of Nunappleton. According to this second reading, which suggests a restricted reformation, only within the protected walls of this particular household can an insubstantial virtue such as virginity exercise any power. The confusion, of course, leaves unresolved a problem with which this chapter has been centrally concerned: the problem of the political efficacy—the agency—of Mary's medial state of virginity, and the

---

45. Harry Berger, Jr., in "Marvell's 'Upon Appleton House': An Interpretation," *Southern Review* (Australia) 1, no. 4 (1965): 16–18, discusses the interpretive options presented by this textual crux in lines 761–62: "The previous stanza (lines 753–60), with its suggestion that the latest is the best, would opt for Baconian optimism, while the political chaos of Marvell's time would encourage chiliastic pessimism." See also Colie, "*My Ecchoing Song,*" p. 264.

generally private, passive, household virtues with which that virginity is aligned. As the audience of nature's woodland masque earlier in the poem, the narrator witnessed the hewel exercise a domestic virtue that seemed, remarkably, to purify Fairfax's wood; the question raised now at the poem's conclusion is whether the political energies depicted in that masque were as vacuous and mystified as the royalist political magic enacted in the Caroline masques at Whitehall. Can this same principle of spiritual reformation work to redeem an entire nation? Can the studious exercise of a domestic virtue such as chastity function to reform the world that lies outside these gardens of repose? The second interpretive option seems to deny the possibility of this historical idealism, postponing the redemption to the end of the temporal world. According to this latter reading, only an extraordinary act such as marriage, or an even more desperate type of sacramental violence, can precipitate the vitrification of that rude heap to the crystalline sphere of apocalyptic perfection.

Marvell's poem concludes with little but ambivalence toward the question of the ultimate efficacy of the self-sufficient power of natural virtue—the vitalist figures of power encoded in the poem as Mary's virginity and Fairfax's chaste conscience. One possible solution to this interpretive knot may be suggested by Marvell's Miltonic source for his figurations of virginity and conscience. One of the most puzzling aspects of Milton's *Mask*, as A. S. P. Woodhouse and Douglas Bush have explained, is "the severe limitation set upon the power of Haemony in the action: if it can protect the Brothers against the enchantments of Comus—against their falling into the same predicament as the Lady—it can do nothing to reverse the spell and rescue her."[46] The natural powers summed up in her virginity are not sufficiently powerful to free the Lady from her chains. The Lady's radical humanist Elder Brother had tried to identify chastity as a "hidden strength / Which ... may be term'd her own" (421–22). But the Second Brother questions this bald Pelagian assertion, and his desire to subordinate the natural strength of chastity to the "strength of Heav'n" (420) is an anxiety the masque bears out: the power of virginity alone—or of whatever natural virtue we wish to see embodied in the Haemony—is simply not strong enough to "undoe the charmed band / Of true Virgin here distrest" (914–15). These natural powers having failed, the masque ultimately concedes that only a divinely vouchsafed power of grace can deliver the Lady from her root-bound stasis: if a natural virtue such as virginity is feeble, the Attendant Spirit concludes, heaven must stoop to its aid. Milton's masque requires ultimately

46. *A Variorum Commentary on the Poems of John Milton*, gen. ed. Merritt Y. Hughes, vol. 2, part 3 (New York: Columbia University Press, 1972), p. 936.

the redemptive agency of an external power, and it is surely pertinent to our understanding of Marvell's reading of Milton that the nymph of grace, Sabrina, derives her transcendent powers from an action of self-sacrifice (839–42). The image of a virgin sacrifice points us toward the Lady's inevitable destiny in the sacrifice of her virginity, as *Comus* concludes with the prospect, albeit an oblique one, of the Lady's translation from the state of virginity to the presumably higher ideal of marriage (1013–21).

If, as I have suggested, Marvell's debt to Milton's *Comus* involves in part an avowal of Milton's ultimate concession to the possible limitations of autonomous human virtue, then we can, perhaps, recognize an unwilling but inevitable circumscription in the country-house poem of the private virtues represented by Marvell as the virginal modesty and innocence rooted in the household. The ethical conclusion Marvell's narrator seems to draw at the close of the poem is twofold: Mary's retired virginity will allow the flourishing of organic process, but the claims of this natural reformation are limited to retirement. Even as he idealizes the promise of a natural regeneration, the narrator seems finally to concede its failure to encompass the troubled world of political strife. While clearly admiring the limp virtue of the retired Lord General and the virginity of his daughter, the speaker ultimately recognizes the inefficacy of this faith in natural providence. The likelihood that a fugitive and cloistered virtue such as virginity is not sufficiently potent to redeem a nation ultimately demands a concession to a faith in the vitrifying forces of a divine power: the home-bred virtues of the private sphere must be forsaken in order to institute a "universal good" that can cross the boundary from the private realm to the public "*World*" (761). In acknowledging this concession to the world of action and necessity, Marvell does not retract his characterization of marriage as sacrifice. Even as it celebrates the eventual marriage of Mary Fairfax and the sphere of public action that that marriage represents, *Upon Appleton House* implicitly laments the loss of the effect of virginity on the garden world of the estate. This poem bemoans the impotence of the individual virtuous soul, and virtuous body, to organize the increasingly chaotic body politic.

4 Chaos, Creation, and the Political Science of *Paradise Lost*

"Milton's Utopia never rose above chaos."
—Arthur Barker, *Milton and the Puritan Dilemma*

Well before Marvell had begun to write the poems that would draw on *Comus*'s affirmation of the bodily power of the virgin, Milton had abandoned his commitment to both the practice of celibacy and the theory of virginal apocalypticism.[1] But when he came in the 1640s to champion the state of holy matrimony, he did not abandon the image of bodily integrity as the figural basis of his liberatory politics and theology. Milton's early interest in the bodily state of virginity has been seen to generate the extraordinary doctrines of animist materialism and mortalism that would come to distinguish the heretical temper of *Christian Doctrine* and *Paradise Lost*. The path from youthful celibate to mature monistic materialist may not at first seem an obvious one. But we might usefully turn to a statement by Denis Saurat, the first critic to explore the radical nature of Milton's materialism, for a clear-sighted, if simplistic, mapping of this path: "Milton was driven to pantheism by his pride and chastity: his body was holy in his eyes; his body will be of the substance of God; matter will be of the substance of God."[2]

During the years of Marvell's stay at Nunappleton, from 1650 to 1652, the period of the most intense activity among vitalist intellectuals, Milton had not yet begun a formal articulation of his monistic materialism. This work he would begin at the end of the decade, in chapter 7 of the first part of *Christian Doctrine* and in *Paradise Lost*. Milton was occupied instead at this time as a propagandist for the republican cause, writing by far the most radical of his political treatises, *The Tenure of Kings and Magistrates* (1649), *Eikonoklastes* (1649), and, after he had been appointed Latin Secretary by

---

1. On the question of the young Milton's investment in the ideal of celibacy, see Leonard, "Saying 'No' to Freud"; and Kerrigan, "Politically Correct *Comus*."

2. Denis Saurat, *Milton: Man and Thinker* (New York: Dial Press, 1925), p. 46. For a different, far more elaborate account of the origin of Milton's materialist heresies, see Kerrigan, "Heretical Milton."

the Council of State, his *Defensio pro populo Anglicano* (1651), or *The Defence of the English People*. Although an early engagement of the monistic union of body and spirit can be detected in Milton's divorce pamphlets (1643–45) and the *Areopagitica* (1644), the explicit purpose of the treatises at midcentury is the theoretical justification of the regicidal origins of republicanism.[3] Despite the necessarily practical function of these three pamphlets, we can discern the impact on the propagandist of the contemporaneous welter of interest in the philosophy of monistic vitalism. As we will see, the short-lived vitalist movement that sought with such urgency to establish a science of self-motion pushed Milton to a new conception not only of material bodies but of the body politic.

The Distribution of Vital Spirit

I hope ultimately here to situate in its political context the relation of philosophical vitalism to the Creation from chaos depicted in *Paradise Lost*. I examine, among other aspects of the poem's politically engaged science, the divergent creation narratives of Books Three and Seven and, in the final section of the chapter, the set-piece chaos through which Satan flies at the end of Book Two. In order to follow the rich conversation in Milton's epic between politics and science, I propose we begin with a look at a curious document from the third year of the English Commonwealth. One of the countless contemporary records of Milton's public activity quarried from the files of Britain's Public Record Office appears in the form of a brief note to the Examinations Committee of the Council of State, entered on March 5, 1651, regarding Milton and the publication of two treatises.[4] The first matter noted in the entry is a recommendation to the committee to reprint one of Milton's recent political tracts, probably the *Defence of the English People*, published first a few weeks earlier. The second item noted on the document is a directive to the same committee to examine a complaint, issued by Milton himself, concerning the unauthorized translation of a Latin text whose patent had been owned by Milton's printer, William Dugard. The unau-

---

3. On the early intimations of Milton's vitalism, see Stephen M. Fallon, "The Metaphysics of Milton's Divorce Tracts," in *Politics, Poetics, and Hermeneutics in Milton's Prose*, ed. David A. Loewenstein and James Grantham Turner (Cambridge: Cambridge University Press, 1990), pp. 69–83. Kendrick discusses the monistic imagery of *Areopagitica* in *Milton*, pp. 21–35.

4. *The Life Records of John Milton*, ed. J. Milton French, 5 vols. (New York: Gordian Press, 1966), 3:6: "Wednesday the 5th of March 1650. . . . That it be referred to y^e Comm^{ttee} of Ex$\bar{\text{i}}$ons to veiw over Mr Miltons booke & to give Order for reprintinge of it as they think fit. And that they also examine the complant by him made about Peter Cole his printing a Copie concerning the Ricketts wch Mr Dugard alledgeth to be his."

thorized publication of which Milton complained is a translation of Francis Glisson's 1650 *De rachitude*, or *A Treatise of the Rickets*, a work Glisson had written in collaboration with the help of Milton's friend, Dr. Nathan Paget.[5] This strange note to the Order Books Committee, written at the height of the poet's influence with the Council of State, has confronted all Milton's modern biographers with the conundrum of connection: what has Milton's *Defence* to do with a technical investigation of rickets? The biographers, not at all surprisingly, have all followed David Masson in dismissing as insignificant the "second part of the entry, so oddly connected with the first."[6] There is, to be sure, very little investment in childhood bone pathologies charging the pages of the *Defensio*, the Latin treatise Milton wrote shortly after the execution of King Charles I to justify the ways of the regicidal English to the incredulous European intellectual community; and there is to my knowledge no evidence of a subversive political drive motivating the physiological studies of the royalist Francis Glisson. I wish to propose, though, that we give at least a provisional consideration of the possibility that there is more than mere coincidence that brings together two texts as generically and thematically distinct as Milton's *Defence of the English People* and Glisson's *Treatise of the Rickets*. A brief examination of Glisson's achievement in his decidedly vitalist treatise on rickets should introduce us to some of the surprising and consequential ways in which the widely dispersed energies of the Vitalist Moment intersect with Milton's own concerns in the period from 1649 to 1651 and which continue, well into the 1660s, to shape the philosophical and ideological contours of Milton's great epic.

Like Harvey's *Circulatio sanguinis*, Glisson's *De rachitude* is a landmark of seventeenth-century English physiology.[7] Far more than a study of the

5. Francis Glisson, *A Treatise of the Rickets, being a Diseas common to Children. Published in Latin by Francis Glisson, George Bate and Ahasuerus Regemorter. Translated by Phil. Armin. . . . Printed by Peter Cole* (London, 1651). All page references to this work are cited in the text. Milton appears to have been concerned that Glisson's treatise be published properly, by the printer in possession of the copyright. According to Webster, *Great Instauration*, this printer, William Dugard, may have gone on to translate a collection of Paracelsian medical writings, *Paracelsus His Dispensatory* (London, 1654).

6. David Masson, *The Life of John Milton*, 5 vols. (London: Macmillan, 1877), 4:313. W. R. Parker, in *Milton: A Biography*, 2 vols. (Oxford: Clarendon, 1968), 2:978, and Hill, in *Milton and the English Revolution*, p. 492, ascribe Milton's complaint to his obliging attempt to support either his publisher Dugard or his friend Paget in a matter in which he himself could have had no intellectual investment.

7. Walter Pagel describes Glisson's radical reconceptualization of physiological organization in *New Light on William Harvey* (Basel: Karger, 1976), pp. 34–36; "The Reaction to Aristotle in Seventeenth-Century Biological Thought," in *From Paracelsus to Van Helmont*, pp. 503–8; and "Harvey and Glisson on Irritability with a Note on Van Helmont," in *From Paracelsus to Van Helmont*, pp. 497–514. See also Thomas S. Hall, *Ideas of Life and Matter:*

specific childhood disease of rickets, Glisson's treatise makes its impact felt through a radical rearticulation of the nature of the substance of the body and, by extension, of the nature of substance in general. As Walter Pagel has observed, it is in the *De rachitude* that we find the first formulation of the principle of tissue irritability, a physical defense mechanism operative in bodily tissue that allows it to respond locally to injurious external stimuli. At the heart of the theory of tissue irritability was the vitalist understanding of bodily substance that had begun to avail itself to English intellectuals in the late 1640s. As if struggling to heal the wound of dualism inflicted by Descartes, Glisson, like his friend William Harvey, continually generates narratives of the cooperation and interpenetration of body and soul, as in this account of the infusion of solid flesh with "Vital Spirit": "Life cannot consist without a Vital Spirit. Therefore when the Vital Spirit is distributed in and with the Arterious Blood to the solid parts through the Arteries, and these parts do suck in that Blood into their substance, it comes to pass, that the said parts are co-united with the Vital Spirits, and so they participate of the Nature of Life" (p. 100). Glisson begins his account with a provisional dualism that distinguishes between the "solid parts" of bodily substance and the energy force of "Vital Spirit." But after a willful sucking of spiritualized blood, these "said Parts," no longer the passive lumps of matter imagined by the mechanists, are forever "co-united with the . . . Nature of Life." In the pages of Glisson's disquisition on the vital tissues of the human body, we have an early adumbration, within the generic parameters of physiological description, of the theological heresy of monism that Milton would codify later in his *De doctrina christiana*. Milton argues in his theological treatise, probably written concomitantly in the late 1650s with *Paradise Lost*, for the same "co-union" of body and spirit. The spirit, or soul, holds no "intelligent existence independently" of the body and is no more mysterious than the vital, sensitive, and rational faculties that provide life and energy to the human animal.[8]

As with Harvey, the focus of Glisson's physiological study is bodily organization, the disposition and distribution of agents and resources within a closed system. Although Glisson appears to have formulated the vitalism of his treatise independent of Harvey's 1649 account of animated, self-moving

---

Studies in the History of General Physiology, 2 vols. (Chicago: University of Chicago Press, 1969), 1:396–97.

8. Milton argues for the identity of human soul and animal vitality in *De doctrina christiana*: "Nor has the word 'spirit' any other meaning in the sacred writings, but that breath of life which we inspire, or the vital, or sensitive, or rational faculty, or some action or affection belonging to those faculties. . . . Hence the word used in Genesis to signify 'soul,' is interpreted by the apostle, I Cor. xv. 45. 'animal' " (15:39–41).

blood, the *De rachitude*, like the *Circulatio sanguinis*, boldly dismantles the essentially monarchic paradigms of authority and organization that had characterized all early modern physiology. The brain, the heart, the stomach— all the traditional centers of bodily control—are dethroned as the agents of the body's government. In their place rise the random, disparate masses of bodily tissue that find themselves capable of actions and reactions independent of any efferent center of command. Bodily health, then, and bodily disease, rest not in what Glisson calls the "public," governmental organs of the brain or the heart; the responsibility for the body's condition lies rather in the "private" members, the limbs and masses of tissues situated more or less along the body's periphery. The condition of rickets, for example, is the result of the same form of inequity faced by the hungry citizens in Menenius Agrippa's fable, the malady on which Glisson bestows the name "Alogotrophy," or the "irrational Nourishment of the parts" (p. 342). The explanatory logic behind Glisson's analysis of rickets is reminiscent of Harvey's "liberal" analysis of the circulatory system in 1649. Although the heart, for Glisson, still maintains control over the movement of the blood, "the first cause of this unequal circulation of the bloud is some disposition of an outward . . . member, laboring under some private Disease" (p. 30). If such a bodily member, privately motivated, finds itself without sufficient sanguineous nourishment, that is because this private member has done little but "sparingly sip that blood," a local, self-determined failing that causes the blood to be "distributed with an unusual liberality thorow the other parts" (p. 87). The Glissonian body maintains a more distinctly monarchic structure than its Harveian counterpart, but this provisionally royalist organization is nonetheless authorized and determined by a liberally structured network of private bodily members and local bodily forces.

It is in reference to the general logic of private responsibility and decentralized determination structuring Glisson's treatise that we can entertain the possibility of a larger significance than mere temporal proximity connecting the two texts brought to the attention of the Council of State on March 5, 1651. Glisson's *De rachitude* and Milton's *Defensio* both sketch the outlines of the radically novel, identifiably liberal structure of organization recognizable in the body described in the same year in William Harvey's *De generatione animalium*. Glisson figures a body of loosely related tissues and members operating without the supervision of a central nervous system, while Milton forwards a Leveller-influenced theory of the body politic whereby a commonwealth of rational men organize themselves in the absence of the centralizing pressures of an absolute monarchy. "Where many are equal," Milton reasons in one of the more populist passages in the *Defence*, "as in all governments the majority are, they ought, I think, to have an equal interest in the government and hold it by terms" (7:127). Like the outward limbs that are granted in the *Treatise of the Rickets* the capacity to negotiate their own intake of common resources, the people in Milton's idealist *Defence* are entitled to govern themselves through the mechanism of individual choice: "all nations and peoples have always possessed free choice to erect what form of government they will" (7:77).[9] The body natural and the body politic are obviously on some level distinct conceptual categories, the theoretical practices describing them inhabiting distinct literary genres. But a sensitivity to the generic disparity between works of natural and political philosophy need not blind us to the shared conceptual logic structuring the systems devised in both these texts, liberal, free-market systems that, as we saw in Chapter 1, had only recently availed themselves to seventeenth-century analytic discourse. It is the abstract logic of the equal allocation of power and responsibility to discrete and only loosely related elements in an organization—implicitly, any organization—that draws together Milton and Glisson as like-minded participants in the Vitalist Moment.

At least since Andrew Marvell, readers of *Paradise Lost* have expressed either relief or disappointment that a political activist such as Milton could emerge from the fray of the Revolution to write an epic from which the consideration of contemporary politics seems to have been so carefully expunged.[10] It is true that Milton's poem exercises a strong resistance to readers who ask poems to unfold their contemporary significances through historical allegory. The sustained speculation of the ideal political organization that had filled so many volumes of Milton's polemical prose does not appear in so recognizable a form in *Paradise Lost*.[11] In Chapter 5 I examine the poetics of political agency inscribed in Michael's treatment of history in Books Eleven and Twelve. Here, though, I focus on the middle books and their inscription of more mediated manifestations of political speculation. The practice of political philosophy had always for Milton involved the engagement of an organizational discourse, a language describing the systemic interaction of individual agents that political philosophy was obliged to share with natural philosophy. I propose that we examine the thousand or so lines

9. Cf. *The Tenure of Kings and Magistrates*, in which Milton explains that the "right of choosing, yea of changing thir own Goverment [*sic*] is by the grant of God himself in the People" (5:14).

10. Mary Ann Radzinowicz discusses the traditional depoliticization of *Paradise Lost* in "The Politics of *Paradise Lost*," in *Politics of Discourse: The Literature and History of Seventeenth-Century England*, ed. Kevin Sharpe and Steven N. Zwicker (Berkeley: University of California Press, 1987), pp. 204–6.

11. For an opposing reading of Milton's poem as a "course in political education," see Radzinowicz, "Politics of *Paradise Lost*," pp. 204–29.

of natural philosophical speculation in Milton's epic—those long sections of Books Five and Seven in which Raphael meditates on the organization of the cosmos—as that literary space that houses the concerns with political organization most pressing to the Restoration Milton.[12] The theorization of chaos and the Creation from the original matter—a practice that would soon acquire the name "chaology"—was for Milton and his contemporaries the study of the most elemental form of material organization imaginable. As such, chaology supplied a controlled and protected discursive laboratory in which it was possible to articulate the metaphysical foundations of the state. It is the chaology of *Paradise Lost* that provides the most legible, if curious, space for Milton's ongoing struggle to articulate a satisfactory ontology of the ideal body politic.

Before we can understand fully the functional entanglement of science and politics in *Paradise Lost*, we must first acknowledge the long decade of political upheaval that separates the author of the *Defence of the English People* from the Restoration poet of *Paradise Lost*. The treatises on the organization of the polity Milton composed between 1649 and 1660 perhaps first and foremost impress the reader with what Arthur Barker rightly described as the "bewildered and disordered reasoning with which [Milton] strove to express his convictions."[13] There may be no pamphlet written in this period in which Milton does not assert contradictory positions on the constitution of the ideal political organization. Despite the welter of often conflicting formulations of the egalitarian or hierarchical arrangement of the ideal polity, a pattern of political philosophical development is nonetheless discernible in these treatises. We have seen the generally liberal impulses behind the idealist philosophy of popular sovereignty marking Milton's *Defence*. Not long after its publication, however, Milton edged away from the faith, expressed (however haltingly) in his treatises of the Vitalist Moment, that God had invested the power of self-determination in the "people."[14] He began instead to forward a more authoritarian vision of a state governed not

12. I do not discuss here the politically resonant astronomy of Book Eight. For an analysis of the social and specifically domestic implications of Book Eight's celestial discourse, see John Guillory, in "From the Superfluous to the Supernumerary: Reading Gender in *Paradise Lost*," in *Soliciting Interpretation: Literary Theory and Seventeenth-Century English Poetry*, ed. Elizabeth D. Harvey and Katharine Eisaman Maus (Chicago: University of Chicago Press, 1990), pp. 68–88.

13. Arthur Barker, *Milton and the Puritan Dilemma, 1641–1660* (Toronto: University of Toronto Press, 1942), p. 217.

14. See Ernest Sirluck, "Milton's Political Thought: The First Cycle," *Modern Philology* 61 (1964): 209–24; and Barbara Kiefer Lewalski, "Milton: Political Beliefs and Polemical Methods, 1659–60," *PMLA* 74 (1959): 191–202, esp. 198–99. See also Annabel Patterson's observations on the *Defensio* in her *Reading between the Lines* (Madison: University of Wisconsin Press, 1993), pp. 232–36.

by the people but by a parliament and, finally, by an even more select and
limited council of the nation's ablest and most rational citizens. The radically
liberal theory of popular sovereignty was transmuted over the course of a
decade into an aristocratic theory of the government of the people (referred
to now as the "masses") by a worthy minority, or revolutionary elite. In 1659,
Milton came even to insist that if such a rational oligarchy were ever established in England, it would be justified in imposing its will on the unruly
majority through the exercise of military force.[15] Within ten years, Milton's
idealist faith in the people's divinely sanctioned political self-determination
hardened into a commitment to the subjection of this same people to a
militarized oligarchy of Puritan "saints."[16]

Given the period's organizational imperative—the contemporary intellectual pressure to formulate a natural philosophy from which a political
philosophy could be derived—we should not be surprised to see a representation of material organization in *Paradise Lost* that accommodates the
conservative changes rung in Milton's theory of the state. The epic's depiction of the original matter should, at some level, retrace the organizational
outlines of Milton's late theory of the polity. It is the poem's primary account
of material organization in *Paradise Lost*, Raphael's beautiful description of
the scale of nature, that provides just this metaphysical justification for the
saintly, if vaguely repressive, rule of the "true commonwealth" charted in
Milton's final treatises:

> O *Adam*, one Almighty is, from whom
> All things proceed, and up to him return,

15. Milton writes, in a passage in *The Readie and Easie Way* that Barker rightly identifies as
"painful," "More just it is doubtless, if it come to force, that a less number compel a greater
number to retain, which can be no wrong to them, their liberty, than that a greater number for
the pleasure of their baseness compel a less most injuriously to be their fellow slaves" (6:141).
See Barker, *Milton and the Puritan Dilemma*, p. 272. Milton reveals himself in 1659 to be no
less resigned than the God of *Paradise Lost* "to subdue / By force, who reason for thir Law
refuse" (6.40–41). Milton's troubled justification of the forceful compulsion of "liberty" surely
merited the sharp criticism of an anonymous wag who rightly noted in *The Censure of the Rota
upon Mr Miltons Book, Entituled, The Ready and Easie way to Establish a Free Common-wealth*
(London, 1660), Milton's ironic proximity to the militant Fifth Monarchists: "the Fift *Monarchy*
Men . . . would have been admirable for your purpose if they had but dream't of a fift Free-
State" (p. 12).

16. Whereas Lewalski, in "Milton," ascribes the contradictions of Milton's political philosophy to a series of pragmatic political accommodations, Barker, in *Milton and the Puritan Dilemma*, more convincingly attributes the "illogicality" of Milton's politics not to a concession to
events but to a sincere intellectual struggle with the problem of social and political organization
(p. xx). I am indebted to Sean O'Sullivan here for his reading of these two critical assessments
of Milton's waffling.

> If not deprav'd from good, created all
> Such to perfection, one first matter all,
> Indu'd with various forms, various degrees
> Of substance, and in things that live, of life;
> But more refin'd, more spiritous, and pure,
> As nearer to him plac't or nearer tending
> Each in thir several active Spheres assign'd,
> Till body up to spirit work, in bounds
> Proportion'd to each kind.
>
> (5.469–79)[17]

Raphael reveals here the gradualist chain of all things that ties the barely spiritualized matter of gross substance to the ethereal, rarefied matter of pure spirit. If we do not see in Raphael's natural philosophy the homogeneous saturation of all matter with spirit that we have come to identify with midcentury vitalism, that is, I suggest, because we do not see in the post-Restoration Milton a consistent faith in a liberal political philosophy derivable from such a vision of matter. Milton had been driven by the late 1650s to reconfigure the political state as a rude multitude governed from above by a "rational" elite that one critic has properly identified as an "aristocracy of grace."[18] Raphael's image of the differentially spiritualized things of nature charts a hierarchical organization identical to the only Puritan commonwealth Milton, by the late 1650s, was able to envision: the crudest elements of the system find themselves hierarchically subjected to the fit few, the beings "more refin'd, more spiritous, and pure" (5.475). Inflexible stratification is as much the focus of Raphael's vision as ontological mobility: the "things" of Raphael's nature, like the citizens of Milton's projected society, are circumscribed by "assign'd" spheres and "proportion'd" bounds. Milton's pained political resignation to a doomed politics of minority rule finds a sanguine cosmological justification in Raphael's hierarchical continuum of body and spirit.

17. All quotations from Milton's poetry, and all line citations, are taken from *John Milton: Complete Poems and Major Prose*, ed. Merritt Y. Hughes (Indianapolis: Odyssey, 1957). The best account of the natural philosophy of this speech is in Fallon, *Milton among the Philosophers*, pp. 102–7. I have been aided as well by the discussions of Milton's monism in Kendrick, *Milton*, pp. 179–205; Harinder Singh Marjara, *Contemplation of Created Things: Science in "Paradise Lost"* (Toronto: University of Toronto Press, 1992), pp. 210–19; and especially Kerrigan, *Sacred Complex*, pp. 193–262.

18. Merritt Y. Hughes, "Milton and the Sense of Glory," *Philological Quarterly* 28 (1949): 119. Fallon has noted perceptively how appropriate it is that "Milton, a revolutionary who distrusted the masses, could find attractive a picture of a scale of homogenous matter crowned by rare spirits," in *Milton among the Philosophers*, p. 109.

The Matter of Revolution

Political circumstance may have forced Milton to discard, by the time he was composing *Paradise Lost*, the egalitarian vision of popular sovereignty that makes the treatises of 1649 and 1651 so daring in their enfranchisement of the *populus*; his poem does not consistently articulate, at least in political philosophical terms, a belief in the equality of human beings. But I hope now to demonstrate that the liberal theory of organization Milton had embraced in the vitalist period is not missing altogether from the poem published in 1667. The egalitarian discourse of the treatises of 1649 and 1651 does, in fact, find voice in the Restoration epic. But this obsolete liberalism does not surface in the later period in the familiar guise of political philosophy. The organizational cover under which Milton's radicalism appears in *Paradise Lost* is an alternative science.

Scholars of Milton's monistic philosophy of matter have accorded Raphael the last word as he weighs for Adam the merit of the "various degrees / Of substance" in the description we have just examined of the scale of nature in Book Five. His speech on the gradations of material quality presents itself as the poem's most formal, and therefore presumably official, statement on the organization of the orders and degrees of substance. But Raphael's hierarchical scale of nature is by no means the only map of material organization Milton draws in the late 1650s and early 1660s. There are moments not only in *Christian Doctrine* but in *Paradise Lost* as well in which Milton represents the generous, egalitarian vitalism he would have found in texts such as Glisson's *De rachitude* and Harvey's *Circulatio sanguinis*, vitalist works in which matter is not segregated by degrees of spiritualization but infused uniformly with spirit and energy. The liberal extension of sovereign power that was no longer viable in the political sphere is displaced onto the sphere of ontology. The theory of popular empowerment that Milton had felt compelled to bury, ten years earlier, in its strictly political avatar reemerges in both *Paradise Lost* and *Christian Doctrine* as a vitalist theory of matter. When Milton describes in these late works the propagation and extension of divine spirit throughout the entire mass of the original matter, we can see a recrudescence of the radical philosophy of organization he had embraced in the great treatises of the vitalist period, *The Tenure of Kings and Magistrates* and *The Defence of the English People*.

It is in Chapter 7 of Book One of *Christian Doctrine* that Milton unfolds the most liberal strains of his natural philosophy, a strain that approaches at times the exuberant egalitarian vitalism of a Glisson or a Winstanley. The human soul for Milton, as for so many of the vitalists, does not govern the body in an incorporeal state of splendid isolation: it is "equally diffused throughout" the whole of the "organic body" (15:49). And it is this egalitarian image of human bodily organization that provides the figural model for Milton's metaphysical representation of the larger body of chaos. In deriving chaos from the actual body of God, Milton renders it, and all its subsequent productions, inherently good: "The original matter of which we speak, is not to be looked upon as an evil or trivial thing, but as intrinsically good, and the chief productive stock of every subsequent good" (15:23). This inherent chaotic goodness becomes the cornerstone of Milton's ambitious theodicy: the only God whose ways Milton would be capable of justifying is a God who was willing even before the Creation to share his goodness and his power with his disidentified body that becomes the matter of chaos. The Deity's generous extension of sovereign virtue beyond the confines of his own essence elicits from Milton this gracious acknowledgment: "It is an argument of supreme power and goodness, that [the] diversified, multiform, and inexhaustible virtue . . . substantially inherent in God . . . should not remain dormant within the Deity, but should be diffused and propagated and extended as far and in such manner as he himself may will" (15:21–23). By means of this effusive praise of a generous Creator, Milton performs the initial move in what we can identify as a massive liberalization of the cosmos. He oversees the dispersal of a power that need not remain hoarded within a sanctuary of systemic control, because it can be more efficiently redistributed to a mass of matter that can safely be trusted to govern itself. By depicting a God willing to diffuse his virtue across the entire tableau of creation, Milton establishes the only theological foundation that can support the hypotheses of vitalists such as Glisson and Harvey, or, for that matter, of free-market philosophers such as Misselden and Mun. Milton decentralizes divinity, representing an action logically prior to the decentralizations of the state, of the economy, and of the human body. In describing in his *Christian Doctrine* the first divestment of kingly power, Milton subsumes into his cosmogony, just as we might expect, all the fathers of modern liberalism.

The representation most faithful in its narrativization of this doctrine of the diffused virtue of chaos is surely that account of Creation related to Satan by Uriel in Book Three. As the Father's chief "Interpretor" (3.656) and the poem's only eyewitness to the Creation, Uriel provides us with what we are clearly to accept as a privileged description of this event. He describes a "formless Mass" of chaos so animated that it "came to a heap" at the sound of the divine voice (3.709–10):

> Swift to their several Quarters hasted then
> The cumbrous Elements, Earth, Flood, Air, Fire,

And this Ethereal quintessence of Heav'n
Flew upward, spirited with various forms,
That roll'd orbicular, and turn'd to Stars
Numberless.

(3.714–19)

After God's initial infusion of divine virtue, the process of Creation proceeds by means of chaos's generative capacity to produce a world, "spirited with various forms," from deep within its sentient bodily mass. The individual elements of earth "haste" to their proper quarters, and the ethereal matter of heaven individuates itself into the numberless stars of the firmament. Uriel, like Raphael after him, incorporates into his scientistic narrative the Genesis formula of the divine fiat (3.709, 712). But as Christopher Kendrick has rightly noted, the creative fiats in *Paradise Lost* are less causal than occasional in nature.[19] The vitalist process of material self-organization is the direct result neither of God's command nor of his vigilant and ongoing manipulation of an inert substance. The process by which "order from disorder sprung" (3.713) was set in motion by an unrepeatable originary act that empowered the world's material mould to order itself; once the abyss has been impregnated with a self-activating *divina virtus*, the effective control over generation devolves on the now self-generating matter of chaos.

Uriel's creation narrative makes clear the organizational implications of the total (as opposed to the hierarchically segregated) spiritualization of the generative matter of the abyss. But we must return to Raphael for the narrative in *Paradise Lost* that pushes in the most radical directions the implications of a truly vitalist Creation. Raphael, we have seen, espoused in Book Five a theory of matter that reflected the conservative turn—documented in the pamphlets of the later 1650s—in Milton's political philosophy. In his story of Creation, however, in Book Seven of the poem, Raphael extends his account of matter to its origins in chaos, developing a chaology that goes far beyond his earlier claims in its radical political suggestiveness. One of the most shocking instances of the complete transmission of divine power to the matter of creation generates an elaborate figural complex that lies at the heart of the political science of Milton's poem: the figure of self-creation. Nowhere does Milton more dramatically wrest from God his creative agency than in Raphael's vitalist account of the appearance of the earth, that event

19. Kendrick, *Milton*, p. 180: "Whereas God's fiats, in Genesis, have the force of imperatives, and work more or less instantaneously, the Son's fiats, in *Paradise Lost*, move away from the imperative and into the exhortative mood." Barbara K. Lewalski discusses the counterpoint, throughout Book Seven, of creation by divine fiat and earthly self-creation, in *"Paradise Lost" and the Rhetoric of Literary Forms* (Princeton: Princeton University Press, 1985), pp. 132–33.

described by Raphael as the emergence of dry land from the water-soaked planet:

> The Earth was form'd, but in the Womb as yet
> Of Waters, Embryon immature involv'd,
> Appear'd not: over all the face of Earth
> Main Ocean flow'd, not idle, but with warm
> Prolific humor soft'ning all her Globe,
> Fermented the great Mother to conceive,
> Satiate with genial moisture, when God said,
> Be gather'd now ye Waters under Heav'n
> Into one place, and let dry Land appear.
>   (7.276–84)

In the final lines of this passage, Milton dutifully reproduces God's commandment in Genesis 1:9: "Let the waters under the heaven be gathered together unto one place, and let the dry land appear." But the figural world of Raphael's narrative account of this event is founded on the authority not of Genesis but of a scriptural text with a very difference ideological valence, the passage from Paul's Epistle to the Romans so important to midcentury vitalists: "For we know that the whole creation has been groaning in labor pain until now" (Rom. 8:23). For Marvell and Winstanley, as we observed in Chapter 2, this verse seemed to guarantee the earth and its naturally perfectionist tendencies a politically significant independence from both human intervention and divine whim. But Milton invokes the Pauline text to authorize an image of autonomy that goes far beyond the protodeism of the vitalist revolutionaries. He pushes the Pauline figure of a parturient earth from the millenarian present all the way back to the actual moment of Creation. Developing a myth of autonomy so radical it borders on the unimaginable, Milton has Raphael describe for Adam the moment at which the earth gave birth to itself.

It is not surprising that such an improbable event should find its rhetorical expression in such a difficult passage of verse. The reason, I have to assume, that these lines have been paid such scant attention by critics involves our difficulty in assigning agency to the central action ("Fermented . . . to conceive") and in separating active verbs ("Appear'd," "flow'd") from verbal adjectives ("involv'd," "Satiate"). But we might profitably view the syntactical infelicities of this passage as a meaningful symptom of the poem's larger struggle with the subject of agency. It is precisely the problem posed by divine action that this passage is laboring to engage and subvert, and we would do well to outline here the specific action Raphael is attempting, however awkwardly, to describe.[20] An "Embryon immature" at first, "in-volv'd," or surrounded, by "Main Ocean," the earth is warmed by the gen-erative humor of this "Womb . . . Of Waters." Softened and made porous, the earthly globe soaks up the saturating powers of the amniotic abyss, ma-turing, through something like endosmosis, into a womb herself. Over the course of five lines, this globe has metamorphosed from "Embryon" to "Great Mother" and through a process of fermentation is made to conceive. It is this sudden and unexpected transformation of a fetal earth to a pregnant Earth Mother that constitutes the mysterious center of the process described by Raphael. What is it that this Great Mother conceives? The answer to this question is at once simple and absurd: the earth conceives and generates the very embryo that was her former self.

The logical problem besetting any claim for self-generation is the obvious fact that the necessary act of parturition presupposes a higher, preexistent *agent* of parturition. The exquisite circularity of Raphael's narrative of the embryonic earth's self-enwombment and eventual self-expulsion is formu-lated clearly to counter this skeptical insistence on preexisting agents. Milton attempts to thwart the seemingly indisputable logic of parturition through a rhetorical trick of temporal inversion: the carefully sequenced images of em-bryonic maturation (277), conception (281), and sexual satiation (282) reverse the real-life trajectory of animal reproduction. It is nothing less than this reversed trajectory that Milton requires to validate the impossible figure of self-creation. The problem of priority at the heart of the chicken-or-the-egg riddle (worried exhaustively in Harvey's *De generatione* of 1651) receives here what we must concede to be a magical solution: the embryonic egg came first because it hatched itself.

The autofertilization of the egg of creation is not Raphael's first oviparous figure for the world's beginning. Echoing the poem's opening invocation, Raphael has before this moment described how "His brooding wings the Spirit of God outspread, / And vital virtue infus'd, and vital warmth / Throughout the fluid Mass" (7.235–37). It was the Deity in those preliminary images that impressed its ambisexual virtue on the distinctly feminine abyss and made it "pregnant" (1.22).[21] Here, however, in Raphael's description of

20. I have found useful Harry Blamires's brief but careful paraphrase of this passage in *Milton's Creation: A Guide through "Paradise Lost"* (London: Methuen, 1971), p. 179.

21. Milton's image of creative impregnation had its sanction in contemporary natural phi-losophy: Walter Charleton asserted that "at the creation God invigorated or impregnated" the particles of matter "with an Internal Energy or Faculty Motive, which may be conceived the First Cause of all Natural Actions or Motions," in his *Physiologica Epicuro-Gassendo-Charltoniana: Or a Fabrick of Science Natural, Upon the Hypothesis of Atoms* (London, 1654), p. 126; quoted in Henry, "Occult Qualities," p. 340.

the appearance of earth, Milton employs a related vocabulary of sexual contact, but he has subjected this action to a significant change: he has removed from the reproductive process all agents of penetration, excluding from the procreative equation even the semimasculine power of the Holy Spirit. Rather than figuring the insemination of earth by an external and implicitly masculine impregnator, Raphael conjures the very differently gendered image of a "Womb . . . Of Waters" that, in an extraordinary usurpation of paternal prerogative, takes it on herself to soak the earth with the seeds of life. Both womb and embryo, Raphael's earth can in no way be said to be "created" or "form'd" by the direct causal mechanism of the divine fiat;[22] we can say only that it expels itself by means of its own parturiency, its uterine "Waters" breaking "with glad precipitance" in a willful heeding of God's exhortation: "Let dry Land appear" (291, 284).

"So absolute she seems / And in herself complete," Adam says in the next book of Eve, making an observation on the phenomenon of female self-containedness to which Raphael (and no doubt Milton) must respond with "contracted brow" (8.547–48, 560). Raphael, famously, takes Adam to task for his imputation of self-sufficiency to Eve (8.561–78). But Adam has merely transferred to the realm of human relations a principle Raphael had already established as foundational to the material cosmos. Milton, of course, has not gone so far in Book Seven as to exclude God from that initial diffusion of power to the vast abyss. But we see him take considerable pains to clear an independent and almost exclusively feminine space of creative power *after* that initial act of impregnation, laboring to distance the medial stages of Creation as far as possible from the imposing arm of the Heavenly Father: the feminine earth had been "form'd" from the start to function efficiently without assistance from paternal authority.[23] We saw in the last chapter the way in which Marvell's *Upon Appleton House* reserved its most extravagant praise of power for Mary Fairfax, the girl whose virginity at least temporarily

22. Jackie DiSalvo has noted rightly of this passage that "God's role in establishing the universe by fiat [is] almost completely overshadowed by Mother Earth's vivid, procreative activity," in *War of Titans: Blake's Critique of Milton and the Politics of Religion* (Pittsburgh: University of Pittsburgh Press, 1983), p. 105.

23. Christine Froula, in "When Eve Reads Milton: Undoing the Canonical Economy," in *Canons*, ed. Robert von Hallberg (Chicago: University of Chicago Press, 1984), p. 166, and Mary Nyquist, in "Gynesis, Genesis, Exegesis, and the Formation of Milton's Eve," in *Cannibals, Witches, and Divorce*, ed. Marjorie Garber (Baltimore: Johns Hopkins University Press, 1987), pp. 198–99, have argued for Milton's ideological investment in the *suppression* of maternal reproductivity. They focus in particular on that "mother-daughter dyad" that Nyquist identifies as "potentially the most threatening social unit to an insecure patriarchal order since it contentedly excludes the male" (p. 198). Far from suppressing unequivocally this image of self-sufficient femininity, however, Milton deploys it as one of his poem's most potent if dangerous figures for the cherished ideal of human freedom in a nonpredestinary universe.

exempted her from the constraints of external control. Milton's image of a self-sufficient femininity, his figure at least since *Comus* of a radical independence from paternal jurisdiction, functions in a closely related way. The focus on a self-sufficient feminine process works to reorient the ethos informing the traditional rhetoric of agency: in avoiding the conventional scientific figuration of the imposition of masculine force upon passive feminine matter, it functions to reconfigure the authoritarian dynamics of power in the world at large.

In this most involuted of Raphael's creation narratives, Milton, I suggest, is drawing on the most radical of the vitalist discourses of the late 1640s. The signal we are given of this retrospective turn to an ideology of ontological equality is the invocation of that process for which many of the vitalists had substituted impregnation as the origin of conception: "Main Ocean . . . with warm / Prolific humor" did not impregnate but *"Fermented* the great Mother to conceive" (7.279–81). The doctrine of the ferment, theorized most completely in the "Image of the Ferment Makes the Mass Pregnant with Semen," from Jean Baptiste van Helmont's 1648 *Ortus Medicinae*, was at the center of a vitalist, alchemical attempt to restructure the relation of agent to patient in the act of natural generation.[24] The century's premier theorist of the ferment (and one of the greatest influences on Francis Glisson), van Helmont argued that the earth was fertilized through an internally motivated process of fermentation, and not through the insertion of seeds by an outside power; as the "original beginning of things," the ferment was a "Power placed in the Earth," by means of which God "hath given to the Earth the virtue of budding from it self."[25] Aristotle, whose position on self-creation remained orthodox for early modern science, had insisted on an inviolable logical distinction between creator and creature: "the thing generating, cannot be a part of the thing generated."[26] But the doctrine of the ferment permitted a vitalist like van Helmont to dismantle the rationale behind Aristotle's rather common-sense claim. Replacing the turn to external agents

24. On van Helmont's theory of fermentation, see Walter Pagel, *Joan Baptista Van Helmont: Reformer of Science and Medicine* (Cambridge: Cambridge University Press, 1982), pp. 79–86, and Audrey B. Davis, *Circulation Physiology and Medical Chemistry in England, 1650–1680* (Lawrence, Kans.: Coronado, 1973), pp. 50–58. Kerrigan discusses the relation of Helmontian ferment to Milton's imagery of digestion, in *Sacred Complex*, p. 221, and Marjara looks briefly at the role of fermentation in Milton's Creation, in *Contemplation of Created Things*, p. 211. See Curry, *Milton's Ontology*, pp. 124–31, for a reading of Milton's adherence in *Paradise Lost* to the alchemical theories of the transmutation of the elements. The persistence of a radical fermentive philosophy in the work of the Restoration figures Thomas Willis and Henry Stubbe is examined by J. R. Jacob, in *Henry Stubbe*, pp. 51–53.

25. Van Helmont, *Oriatrike*, p. 31.

26. Quoted in van Helmont, *Oriatrike*, p. 32.

with the inward principle of ferment, van Helmont was able to argue that "there is no outward . . . thing generating: but the seminal lump it self, or the generative Seed, doth keep in it self all things which it hath need of for the managing of generation."[27] By locating in the procreatrix a "seminal lump" that "doth keep in it self all things which it hath need of," the doctrine of the ferment swept away the necessity of outside impregnators and created an image of vital matter as self-sufficient, self-moving, impregnable.

Two years after the appearance of his *Circulatio sanguinis*, Harvey brought out his theory of reproduction, *De generatione animalium*, printed in March 1651 by William Dugard, the printer of both Milton and Glisson. Expanding on the famous motto on his title page, *"ex ovo omnia,"* the vitalist Harvey postulated, quite boldly, that the female ovum was not passive but active, and required no direct contact with male sperm for conception.[28] Harvey's theory of a female-centered generation, aided by a commitment to fermentive as opposed to impregnate origins, was greeted by his learned contemporaries with derision and dismissal.[29] But I suspect that Harvey was willing to risk his reputation with the thesis of a nearly autonomous female conception out of a need, recognizable in many of the period's divergent vitalists, to isolate with the authority of science multiple loci of spontaneous generation, figurations of reproduction that place the agency of development in the developing form itself, rather than in higher reproductive agents.[30] Milton shared with Harvey, van Helmont, and Glisson a commitment to the larger ontological principle of material self-determination. And though Milton's stated interests in self-determination were framed within the discourses

27. Van Helmont, *Oriatrike*, p. 32.

28. The orthodox account of female reproductive passivity is voiced, for example, by the sixteenth-century physician Ambroise Paré: "the seede of the male being cast and received into the wombe, is accounted the principall and efficient cause, but the seede of the female is reputed the subjacent matter, or the matter whereon it worketh" (*The Workes of that Famous Chirurgian Ambrose Parey*, trans. Thomas Johnson [London: Thomas Cotes, 1634], p. 885). For an account of Harvey's elevation of "the *ovum* [to] a position of much higher dignity than it had enjoyed in Aristotle's view," see Pagel, *New Light on Harvey*, pp. 25–27. It goes perhaps without saying that in these observations on the Harveian element of Milton's natural philosophy I take issue with Don M. Wolfe's unyielding pronouncement, "No man of the seventeenth century was more remote from Milton's outlook than William Harvey" (Wolfe, ed., *Complete Prose Works of Milton*, 4[1]:80).

29. Geoffrey Keynes, *A Bibliography of the Writings of Dr. William Harvey, 1578–1657*, 2d ed. (Cambridge: Cambridge University Press, 1953), p. 50.

30. Of Harvey's *De generatione*, Merchant, in *Death of Nature*, p. 158, explains that "Harvey's work on generation led him to differ from Aristotle in assigning an efficient cause to the mother hen, but this did not imply the primacy of the female line in procreation. The hen's egg developed through its own internal principle, its vegetative soul. This soul was not derived from the mother hen, and the egg did not live by the vital principle of the mother . . . but was 'independent even from its first appearance.' Free and unconnected to the uterus, the egg perfected itself by its internal formative force."

of theology and political philosophy, his vitalist expressions of natural philosophy in *Paradise Lost* should be seen as his ideologically informed engagement of these most daring formulations of autonomous generation from the vitalist period. We might reasonably assume that the scornful questioning to which Harvey was subjected by the conservative intellectual Alexander Ross could be with equal justice applied to Milton's Raphael, in his description of the parturient earth: "If this be true," asked Ross, "that the female can thus conceive and generate, what need was there of the Male?"[31] Ross's ruffled incredulity in the face of Harveian conception points to the conservative responses to which Milton's science is susceptible. As Ross implies, the thesis of a self-sufficient and implicitly gynocentric generation brings with it a host of ideological implications: it seems to render insignificant the entire domain of masculinity. The belief in autonomous material generation could be seen further, I suggest, to pose a threat not only to maleness but more consequently to authoritative and authoritarian rule in general. It was Harvey himself who hinted at the larger cultural implications of his thesis of the noncontactual fertilization of the female ovum: he distinguished this new theory of generation from any doctrine that represented conception through the analogy "of a King in his dominions, where his command is everywhere a law."[32] If we take into consideration the wider purview of this antiauthoritarian turn, we can imagine Ross's question being reformulated thus: If individuals can thus conceive and generate their own polity, what need is there of a king?

The doctrine of the ferment, and its corollary theory of autonomous generation, was surely that philosophical position most radical in its unraveling of the authoritarian logic of early modern science. We need look for an example no further than Harvey, who had relied on the fermentive hypothesis to justify the blood's daring usurpation of the monarchy of the heart. In the absence of an image of the impulsion of blood by the heart, how was Harvey to explain the mysterious phenomenon of the blood's self-motion? By what agency could the blood actually move itself? In the 1651 *De gener-*

---

31. Alexander Ross, *Arcana Microcosmi: or, The hid Secrets of Man's Body discovered; In an Anatomical Duel between Aristotle and Galen concerning the Parts thereof: As also, By a Discovery of the strange and marveilous Diseases, Symptomes & Accidents of Man's Body. With a Refutation of . . . Doctor Harvy's Book De Generatione. . . .* (London, 1652), p. 223. Ross had explained earlier in the text: "That there is no active seed in the female for generation, but that she is meerly passive, in furnishing only the Matter or Menstruous bloud with the place of conception, is according to *Aristotle* manifest; because if the female seed were active, she may conceive of her self without the help of the male, seeing she hath an active and a passive principle, to wit, seed and bloud; and where these principles are, there will be action and passion" (p. 28).

32. Harvey, *Anatomical Exercitations*, p. 256.

atione, Harvey conjectures that it is an act of fermentive heat that motivates
the revolution of the blood: "For as in *Milke* set upon the *fire*, and in *Beere*,
we see dayly a *Fermentation*, working, or Intumescence; so is it in the *pulse*
of the *Heart*, in which the *blood*, as by a kind of *fermentation* working up,
is *distended*, and then ebbs, or falls down againe."[33] Proceeding without
recourse to an *"external agent,"* without so much as a pulse from the "monarchy of the heart," the circulation finds its radical origin in the natural, self-
regulating process of fermentation "that is accomplished in the *blood*, by its
own *internal heat*, or *innate spirit*."[34] According to Harvey, the blood qualified for its status as the self-authorizing agent of the circulation by means
of its radical independence from higher agents of generation. The blood, for
Harvey, is the first begotten particle in the animal, but this particle, freed
from its begetter, "is like a Son set free, and dwelling in a mannour of his
own, and a principle subsisting of himself; whence afterwards . . . no part is
its own parent, but when it is once begotten, doth provide for it self."[35]

Given the potentially unlimited antiauthoritarian logic of the principle of
the ferment, we should not be surprised to see it employed by intellectuals
far more radical in their politics than William Harvey, and more radical, in
fact, than John Milton. Indeed, as Mendelsohn has observed, "during the
Republic the chymical-ferment theory of the Creation was seen by its detractors to lie behind a communistic social experiment."[36] In 1657, a treatise
on Helmontian vitalism was dedicated to the regicide Robert Tichborne.[37]
This increasingly entrenched association led the Presbyterian Thomas Hall
to attack the vitalist and alchemical turn of mind as a "Familistical-Levelling-
Magical temper."[38] The science of fermentive creation offered to the more
credulous vitalists at midcentury the promise of a metaphysical justification
for political equality and self-determination. There proved, at least for some,

---

33. Ibid., p. 276. Harvey also relies on ferment, or "leaven," in *De circulatione*: "I think the
first cause of distention is innate heat in the blood it self, which (like leaven) by little and little
attenuated and swelling, is the last thing is extinct in the creature . . . nor is caus'd by any external
agent, but by the regulating of Nature, an internal principle" (*Anatomical Exercises*, ed. Keynes,
p. 187).

34. Harvey, *Anatomical Exercitations*, p. 276.

35. Ibid., p. 273.

36. Mendelsohn, "Alchemy and Politics," p. 40. Mendelsohn distinguishes the political valences of alchemical thought during the 1650s from its appearance in the next decade, during
which, he argues, there was little "connection between the chymical Genesis and political radicalism" (p. 40). I would argue, contra Mendelsohn, that the taint of political radicalism continues to color (though not determine) the perception of vitalism, even among its Royalist
proponents, well into the Restoration.

37. Mendelsohn, "Alchemy and Politics," p. 40.

38. Thomas Hall, *Histrio-Mastix* (1654), published with *Vindiciae Literarum* (London, 1655),
p. 199.

to be a fairly simple logical jump between the science of self-creation and
the more specifically political aspiration, such as this of communist Winstanley, that "every one shall look upon each other as equall in the Creation;
every man indeed being a parfect Creation of himself."[39] Raphael, it would
seem, in describing the fermentive origins of terrestrial conception and parturition, is underwriting a politics far more liberatory in scope than the compromised egalitarianism of the most radical of Milton's political pamphlets.

Strange Point and New

Having examined the radical expressions of vitalist agency in Raphael's
account of Creation, we are in a position now to understand the function of
his natural philosophy as a political discourse within the poem itself. A philosophy of self-creation lies at the heart of the poem's most striking and
consequential bid for political authority. I am referring, of course, to the
natural philosophical justification Satan articulates for his rebellion against
the Father, his claim, formulated in a debate with Abdiel before the outbreak
of the war in heaven, that the angels were not created but had given birth
to themselves by means of their "own quick'ning power." It is the "strange
point and new" that the angels were not only "form'd" by God "such as he
pleas'd" (5.824–25) but formed by God through the agency of his Son
(5.835–38) which elicits what is surely the century's most powerful gesture
of natural philosophical speculation:

> That we were form'd then say'st thou? and the work
> Of secondary hands, by task transferr'd
> From Father to his Son? strange point and new!
> Doctrine which we would know whence learnt: who saw
> When this creation was? remember'st thou
> Thy making, while the Maker gave thee being?
> We know no time when we were not as now;
> Know none before us, self-begot, self-rais'd
> By our own quick'ning power, when fatal course
> Had circl'd his full Orb, the birth mature
> Of this our native Heav'n, Ethereal Sons.
>
> (5.853–63)

Both pro- and anti-Satan Miltonists have agreed that Satan's claim of self-
begetting is central to any understanding of the events of the poem. Ob-

---

39. Winstanley, *New Law of Righteousness*, in *Works of Winstanley*, p. 159.

serving that Satan's heresy "is at the root of his whole predicament," the conservative C. S. Lewis found on this matter an unlikely comrade in William Empson, who stated that the assertion of self-begetting "is the steel armature without which the statue of Satan could not tower so high and danger-ously."[40] Beyond, however, the general acknowledgment that the argument of self-generation is quintessentially Satanic, Miltonists have been curiously silent on the actual content of these lines. They have preferred, rather, to nod assent to Lewis's clever dismissal of Satan's invention of a "theory that he sprouted from the soil like a vegetable": "Thus, in twenty lines, the being too proud to admit derivation from God, has come to rejoice in believing that he 'just grew' like Topsy or a turnip."[41]

Satan's theory of angelic beginning deserves more, I think, than Lewis's affable smugness. As a natural philosophical claim about the origin of a spe-cies—that of the heavenly angels—it demands to be examined in the context of the poem's natural philosophy. In the strictest theological sense, Abdiel is of course right to insist that not only were the angels created, but they were created through the agency of the Son, if we can accept the Son in the role of that originary agent who infused chaos with vital virtue. From this rigid theological perspective, focused exclusively on first causes, Abdiel's in-sistence on angelic createdness should arouse in us little more than an Emp-sonian irritation with a God who fails to share essential information with all his creatures. Satan, on the other hand, is clearly wrong to protest on the basis of a vulgar empiricism that the angels were not created at all: even the most radical vitalist creationism must imagine some originary act of material empowerment. But however faulty the motive and initial premise of his claim, Satan, in his rhetorical amplification of his blasphemy, articulates a physiological process that resembles, in too many respects to ignore, the dynamics of autonomous self-organization that is Miltonic creation.

Abdiel, and with him nearly every critic of Milton's poem, expresses con-sternation at Satan's response to a pious declaration of createdness. John Steadman has gone so far as to doubt that Satan himself "really regards himself as 'self-begot.' "[42] Given the importance of this scene, though, it is incumbent

40. C. S. Lewis, *A Preface to "Paradise Lost"* (London: Oxford University Press, 1942), p. 97; and William Empson, *Milton's God* (London: Chatto and Windus, 1965), p. 84. More re-cently, Regina M. Schwartz, in *Remembering and Repeating: On Milton's Theology and Poetics* (Chicago: University of Chicago Press, 1988), has located Satan's claim at the heart of the poem's representation of evil: "If we were to isolate a single moment as Satan's fall—although his degeneration is more accurately spoken of as a process—it would be here. . . . In a real sense, the entire epic constitutes an extended refutation of Satan's heresy of self-begetting" (pp. 21–22).

41. *Preface to "Paradise Lost,"* p. 98.

42. John Steadman, *Milton's Epic Characters: Image and Idol* (Chapel Hill: University of

on us to determine as precisely as possible what it is in Satan's speech that actually counters Milton's animist materialism. The doctrinal error in Satan's account is certainly his Stoic ascription of the birth moment to fate, that moment at which "fatal course / Had circl'd his full Orb."[43] Satan is clearly struggling to deny God's agency even in that initial moment of divine infusion, the point at which God injected vital spirit into the matter of abyss, enabling the autonomous process of self-raising to proceed of its own energy. But to acknowledge God as an *ultimate* cause behind creation is not to accede without question to Abdiel's insufficiently elaborated claim that the angels were "form'd" by God. Satan, it should now be clear, is not speaking unreasonably to argue that the angels were "self-begot, self-rais'd" by their own "quick'ning power." In the context of the most radically vitalist passages of the *Christian Doctrine* and *Paradise Lost*, arguments in which Milton attributes an almost Glissonian capacity for sentience and perception to material substance, the claim that the angels "begat" themselves by their own quickening power should not have an unfamiliar ring. What *is* "self-raising," if not that process of heavenly creation related to Satan by the vitalist Uriel, for whom "the Ethereal quintessence of Heav'n / Flew upward, spirited with various forms . . . and turn'd to Stars" (3.716–18), the same stars employed throughout the poem as astronomical manifestations of angels? What is "self-begetting," if not Satan's admittedly arrogant term for a process that Raphael, in his description of the earth's fermentive emergence from the amniotic waters, has already described? To "raise," in the seventeenth century as now, was to cause bread to rise through fermentation, or yeast (*OED* III.31.b). To "quicken" was "to reach the stage of pregnancy at which the child shows signs of life" (*OED* II.6.b). With these verbs of fermentation and gestation, Satan places the angels in precisely the position in which we found the self-quickening earth in Raphael's account of the appearance of dry land: a lump of sentient, percipient matter, enwombing its own embryonic self and giving birth.

North Carolina Press, 1968), p. 165. The few sympathetic readings of Satan's claim include A. J. A. Waldock, *"Paradise Lost" and Its Critics* (Gloucester, Mass.: Peter Smith, 1959), p. 71; Robert M. Adams, "A Little Look into Chaos," in *Illustrious Evidence: Approaches to English Literature of the Early Seventeenth Century*, ed. Earl Miner (Berkeley: University of California Press, 1975), pp. 78–79; Kendrick, *Milton*, p. 175; and Empson, *Milton's God*. Empson remarks that "Milton, being so learned a man, would not seriously despise Satan for disbelieving his creation" (p. 88). For scientific support for Satan's theory, in fact, Milton would have had to turn to no less preeminent an authority than William Harvey, who never fully relinquished the doctrine of spontaneous generation. See Arthur William Meyer, *An Analysis of the "De Generatione Animalium" of William Harvey* (Stanford: Stanford University Press, 1936), pp. 45–55.

43. Milton argues in the *Christian Doctrine* that "fate can be nothing but a divine decree emanating from some almighty power" (*Works* 14:27); see also *Paradise Lost* 7.172–73.

Satan has felt his way, for the reasons perhaps of ambition and revenge, to an intuitive grasp of the vitalist doctrine of fermentation, a chemical process of self-motivated quickening that seems as likely as any to have been responsible for the formation of the freely willing, self-determining angels. As Satan no doubt understands, a fermentive, as opposed to an impregnate, beginning guarantees at the level of metaphysics the autonomy of the will, a will liberated from the debt immense of direct filiation to a potentially predestinary Father. Technically, to be sure, we must adjudge Abdiel right that the angels were "form'd," and Satan wrong that this formation was at the whim of "fate." But a reasoned conjecture concerning the more difficult problem of the exact process of angelic generation obliges us to synthesize the reductive theses of both the arguing angels: the seemingly irreconcilable statements that the angels were created by God and that the angels in some way emerged from the matter of heaven can easily coexist in a vitalist scheme of natural creation.[44]

Raphael, as we have seen, provided in his account of the appearance of dry land a representation of the general principle of self-begetting; later in his narrative, we can detect a slightly more encrypted representation of a specific act of self-begetting, the birth of Satan himself. In another of the poem's scandalous expressions of creaturely self-generation, Raphael continues his description of the parturient earth, which, on the sixth day of Creation, "Op'ning her fertile Womb teem'd at a Birth / Innumerous living Creatures, perfet forms" (7.454–56):

> The grassy Clods now Calv'd, now half appear'd
> The Tawny Lion, pawing to get free
> His hinder parts, then springs as broke from Bonds,
> And Rampant shakes his Brinded mane.
> (7.463–66)

It is not difficult to imagine that it was in part the sensuality of the lion's liberation of his "hinder parts" that led Coleridge to single out these lines as "wholly unworthy . . . of the enlarged powers of the poet."[45] Coleridge may have been even more offended, though, by the threatening proximity of this official description of Creation to the Satanic claim of self-raising. Milton's

---

44. Henry More voices the orthodox argument that "only those who deny God and all incorporeal substance; strive to see the origin of motion and all life which gleams in the universe in matter itself," in *Opera Omnia*, 2 vols. (London, 1679), 1:607; quoted in Henry, "Medicine and Pneumatology," p. 31. The vitalist Milton, however, in denying only incorporeal substance, was able to affirm God as he sought the origin of life in the matter of creation.

45. Quoted in Hughes, ed., *Complete Poems and Major Prose*, p. 357.

poem has already observed the scriptural association of the devil and "a roaring lion" (1 Pet. 5:8), in the narrator's description of Satan's first viewing of Adam and Eve: "about them round / A Lion now he stalks with fiery glare" (4.401–2). The "Tawny Lion" here in Book Seven, shaking his brinded mane, also looks ahead to the strangely leonine serpent (with its "hairy mane terrific") that will soon be inhabited by Satan. The first lion of creation, caught in a moment of self-liberation as vital and indecorous as any in the epic, is enmeshed in a network of Satanic association that positions this sanctioned act of self-birthing as a shadowy version of Satan's own. From the depths of the vitalist ontology of creation, we see the hesitant emergence of the Satanic ideology of self-possession and self-authorization, an affirmation of creaturely autonomy from which the poem, in its more theological register, will be forced to distance itself.

The subject of the mechanics of angelic creation raised by Abdiel is by no means marginal to the story Milton is telling. It could even be argued that only once we understand angelic origin are we able to evaluate the inherent organizational order of Milton's heaven and evaluate thereby the competing arguments for the intrinsically egalitarian or intrinsically hierarchical structure of heavenly matter. Do all those ranks and degrees in heaven reflect an actual material differentiation in angelic excellence, or has God merely exalted arbitrarily some angels above others? This question, like the mystery of the Father's begetting of the Son on the day of his exaltation (5.603–4), broods like the Holy Spirit over all the representations of celestial polity in the poem. The seminal importance of angelic origin is not lost on Milton in the *Christian Doctrine*, where he raises the subject of the creation of Heaven and the angels only to advise caution: he insists with an uncharacteristic modesty that he cannot even "ventur[e] to determine anything certain on such a subject" (15:31). In *Paradise Lost*, however, Milton does venture such a determination. Within the context of what is arguably the most blasphemous assertion in the entire poem, he supplies us, in the encoded form of Satan's heretical outburst, with a legitimate description of an origin that the poem must, but cannot, narrate.

As we might infer from Milton's nervous sidestepping of the issue in *Christian Doctrine*, the consequences of the question of angelic origin (and the question implicitly of any creature's origin) are considerable. And the proper frame of these consequences, for the participants in the English Revolution no less than the characters in the poem, is political. Nowhere can the claim of self-creation be seen so starkly as a politically valent argument as when Satan tells the assembled angels that he knows they will choose to cast off the yoke of God, "if ye know yourselves / Natives and Sons of Heav'n possest before / By none" (5.789–91). The dissemination of a new natural philosophy of origin has the capacity in this poem, at least according to Satan, to generate and maintain a revolutionary political program. In order to begin an assessment of the political implications of Satan's nativist philosophy, I wish first to turn to Empson, who understood more clearly than any critic of Milton the political force of Satan's vitalism: "He is talking standard republican theory, and in effect Milton presents that as inherently based, not indeed upon atheism, but on a non-authoritarian view of God as immanent. Satan cannot express it well, because the only God he knows is an authoritarian one whom he considers false; but 'Sons of Heav'n' is at least a metaphorical claim to have an immanent Parent."[46] The God against whom Satan chiefly rails is the arbitrary, voluntarist God of Calvin, the deity Milton devotes so much of *Paradise Lost* and *Christian Doctrine* to arguing out of existence.[47] In the debate with Abdiel, however, Satan's bid to be a Son not of God but of Heaven approaches Milton's own vitalist view of divinity, which is itself based, as we examine in greater depth in the next chapter, on what Empson names a "non-authoritarian view of God as immanent."

Like so many of the radical vitalist pronouncements in Milton's epic, Satan's claim that the angels are self-raised sons of Heaven can be traced to the spirited intellectual ferment of the Vitalist Moment. But the specific origin of Satan's claim of heavenly sonship rests not in the exclusively scientific domain of Glisson, van Helmont, or Harvey. The dominant precursor here, as Empson hinted, is Milton himself, who in the exchange between Satan and Abdiel is looking back at the republican theory of his own radical past. He is looking specifically, I suggest, to his acrimonious exchange with "Salmasius," the French monarchist Claude de Saumaise, whose *Defensio Regia* is the specific text to which Milton's *Defensio Populi* responds. When Satan names himself a "Son of Heaven," he is reproducing the radical redefinition of authoritative sonship that Milton himself had attempted in the *Defensio*, where he struggled to counter Salmasius's aristocratic denunciation of the English Commonwealth's new governing body, the Council of State. Salmasius had risen to an impressive height of polemical scorn as he argued that the inherent ignobility of the republican council members implied their inability to govern the nation; he found preposterous that the Puritan leaders, the "sons of the soil, persons scarce of the nobility at home, scarce known to their own countrymen, should have believed themselves entitled to do

---

46. Empson, *Milton's God*, p. 75.

47. Kerrigan, *Sacred Complex*, p. 255, rightly observes that "absolute voluntarism, a divine will of power alone, is the theological position given to Satan in the poem. . . . In its ultimate good sense this poem attacks the severe voluntarism that some critics believe Milton to have espoused." See John Peter Rumrich, "Uninventing Milton," *Modern Philology* 87 (1990): 249–65.

such things" (7:33). It was Salmasius's aristocratic objection to the ignoble council's "sons of the soil" that elicited from Milton one of the most striking animadversions in the treatise: "Of those whom you call 'scarce noble,' some are second to none of your land or kind in nobility; others, being as it were their own ancestors [*ex se natos*], tread the path to true nobility by way of industry and personal worth, and are comparable with any the noblest soever. They had rather be called 'sons of the soil,' too (it being their own soil!)" (7:33). A few of the Commonwealth's leaders are, as Milton avers, titled members of the English aristocracy. But most of the Puritan saints, like the commoner Milton himself, need not bother fabricating fictions of aristocratic privilege. Rather than sons of the nobility, they are quite simply self-begot ("*ex se natos*"); they embrace with pride the title that names their immanent parentage, "sons of the soil."

We are admittedly surprised to hear in Milton even a figurative claim for terrestrial ontogeny more properly uttered by Gerrard Winstanley. Like the beastly villagers in Marvell's country-house poem, Milton appears in this passage in the *Defence* to raze the social landscape of England to that "naked equal Flat, / Which *Levellers* take Pattern at" (441–42). To proclaim oneself a self-begotten son of the soil is surely to announce one's own ontological equality with all the soil's creatures, an implicit argument for universal enfranchisement that Milton elsewhere in this period voices only with hesitation. But it is just this earlier figuration for an ontologically founded egalitarianism to which Milton returns in the debate between Abdiel and Satan in *Paradise Lost*. As a self-proclaimed "Son of Heaven," Satan seems necessarily to forward a vitalist argument for equality, as the homogeneously spiritualized soil of heaven cannot, we assume, dispose itself into the byzantine order of ranks and degrees that mark the heaven typically described in Milton's poem.[48] And it is, to be sure, an argument for angelic equality that Abdiel hears in Satan's speech, as he scoffs at Satan's vulgarly egalitarian faith in "all angelic nature joined in one, / Equal to him begotten Son" (5.834–35).

The egalitarian implications of Satan's rejoinder, however, seem to be at odds with his claims not only elsewhere in this poem but elsewhere in this very debate. There is a point in the exchange with Abdiel at which Satan's commitment to a philosophy of angelic equality, a philosophy whose figuration is drawn specifically from Milton's own *Defence*, shows signs of fal-

---

48. The vitalist theory of generation, for Harvey in *Anatomical Exercitations*, p. 51, requires a belief in a homogeneous, rather than hierarchically composite, matter: "out of the same *White* of the *Egge* (which all men take to be a *similar body*, and without diversity of *Parts*) all and every the parts of a *Chicken*, whether they be *Bones*, *Clawes*, *Feathers*, *Flesh*, or whatever else, are procreated and fed."

tering. The angels, Satan explains haltingly to Abdiel, are, "if not equal all, yet free, / Equally free; for Orders and Degrees / Jar not with liberty, but well consist" (5.791–93). As we might surmise from the number of instances in which critics have cited these lines (unattributed to their speaker) as expressive of Milton's own faith in the inherently hierarchical organization of the cosmos, Satan here is placed in the odd position of reproducing the official hierarchalism of the poem. He sounds in fact indistinguishable at this moment from the angel he is debating, or the narrating angel, Raphael, who has already, as we have seen, professed the hierarchical scale of nature. Satan of course has an understandable motive for his ontological inconsistencies: in his argument against the priority of the Son, it behooves him to champion an egalitarian heaven; whereas in his assertion of his own priority as "of the first, / If not the first Arch-Angel" (5.659–60), he has a vested interest in maintaining the natural foundation for social hierarchy. The confusion in Satan's logic can be, and no doubt has been, attributed from a characterological perspective to the general condition of Satanic mendacity. But the logical fissure in Satan's reasoning points well beyond the circumscribed issue of his failing as a moral being. Satan's damning self-contradiction in his debate with Abdiel works rather to bring into relief the systematic lapses in coherence in the organizational discourse underlying all of Milton's poem.

We have already witnessed the conflicting organizational principles arise from the various accounts of chaos and creation Milton provides us: the all-encompassing animist materialism asserted in the theological treatise and in Uriel's account of the Creation in Book Three sorts ill with the far more hierarchical vision, relayed by Raphael in Books Five and Seven of the poem, of a scale of nature that situates rarefied over gross matter. As we see in the final section of this chapter, both these representations are at odds with the violent, atomistic account of chaos with which Milton concluded Book Two. The uncertain constitution of matter in this period reflects uncertainties concerning the constitution of a wide variety of structures and organizations, the structure of overwhelming concern in this period being that of the political state. The successive conservative shifts in Milton's political utopianism can be charted rather easily in a narrative account of his two decades of political pamphleteering. But Milton's Restoration poem, which incorporates at one moment and in one literary work an entire history of revolutionary allegiances and utopian dreams, presents itself as a privileged text for our examination of the difficult dynamics of organizational discourse. We see below how the contradictory representations of matter in Milton's poem are symptomatic of the larger problem besetting the very possibility of a common-wealth of Puritan saints.

The Tragedy of Tartar

We have up to this point examined the implications of the multiple figurations of the animist materialism in *Paradise Lost*. We have not, however, considered the troubling fact that the poem's first sustained act of natural philosophical description features a material constitution that falls nowhere on the spectrum of contemporary monistic philosophies. The elaborate set-piece at the end of Book Two of *Paradise Lost* reveals a chaos considerably at odds with the divinely embodied abyss described in the *Christian Doctrine* or in the divergent vitalisms of Uriel and Raphael.[49] The personified Chaos that Satan encounters in Book Two is rebellious and anarchic, and the particulate matter of chaos itself bears none of the seeds of the virtuous, rational self-determination that Milton's theology would seem to demand. The abyss through which Satan flies is that place "where eldest *Night* / And *Chaos* . . . hold"

> Eternal Anarchy, amidst the noise
> Of endless wars, and by confusion stand.
> For hot, cold, moist, and dry, four Champions fierce
> Strive here for Maistry, and to Battle bring
> Thir embryon Atoms; they around the flag
> Of each his Faction, in thir several Clans,
> Light-arm'd or heavy, sharp, smooth, swift or slow,
> Swarm populous.
>
> (2.896–903)

The chaos Milton presents in Book Two is about as far from his or Glisson's monism as the range of seventeenth-century natural philosophy allows. The distinctly heterogeneous structure of chaos in Book Two is atomist; and as with all forms of seventeenth-century atomism, the substance of those indivisible material particles is inert and lifeless, and their motion is effected through a mechanics of impact, or the violent collision of contiguous bodies. Most consequentially, its representation seems to affirm the organizational thesis that posed the greatest threat to Milton's radically humanist vitalism.

Critics and editors have traditionally traced the images in this passage to

---

49. The "clear difference between the account of primitive matter in the treatise and the description of the Deep and of Chaos in the poem" was first noted by Arthur Sewell, *A Study in Milton's Christian Doctrine* (Oxford: Oxford University Press, 1939), p. 127. The most recent and rigorous examination of this discrepancy can be found in Schwartz, *Remembering and Repeating*, pp. 8–39.

their undisputed classical sources in Hesiod, Aristotle, Lucretius, and Ovid.[50] But it should also be possible to situate this particularly bellicose representation of chaos amid some of the clamorous contemporary formulations of the constitution of the ideal political state. Stephen Fallon and Harinder Singh Marjara have discussed the relation of Miltonic vitalism to Hobbesian mechanism, and, to be sure, we can justifiably extend their insights to claim that the spiritless collision of atoms in this passage has been crafted to engage Hobbes's mechanistic materialism.[51] But the connection between Milton and Hobbes cannot fully be understood until we realize that it is not so much Hobbesian science that motivates the incoherent chaos of Book Two as it is Hobbesian politics. The violent collisions of the warring elements here evoke nothing so much as the atomized structure of the originary Hobbesian polity, the state of the war of every man against every man articulated in the famous thirteenth chapter of the first book of *Leviathan*.

The speculation of an atomistically structured chaos played a special role in the later-seventeenth-century practice of organizational speculation. The theorization of a specifically atomistic abyss permitted the articulation of the relative merits of a liberally organized system of individual, perhaps even equal, elements or particles. It functioned, often explicitly, to ground the philosophical speculation about the promise, or threat, of free-trade economics and democracy, those nonauthoritarian systems of economic and political organization to which the adjective "chaotic" was most often applied.[52] It is,

---

50. The classic readings of Milton's theory of matter and his representation of chaos are Hunter, "Milton's Materialistic Life Principle" and "Milton's Power of Matter"; Curry, *Milton's Ontology*; A. B. Chambers, "Chaos in *Paradise Lost*," *Journal of the History of Ideas* 24 (1963): 55–84; and Reesing, "Materiality of God." David Quint, in *Epic and Empire: Politics and Generic Form from Virgil to Milton* (Princeton: Princeton University Press, 1993), p. 274, argues for the influence on this passage in Book Two of Phineus Fletcher's *The Apollyonists*.

51. See Fallon, *Milton among the Philosophers*, pp. 126–30; and Marjara, *Contemplation of Created Things*, pp. 89–102.

52. Edmund Waller, for example, associates the Lucretian chaos with democracy in his poem prefacing John Evelyn's translation of Book One of *De rerum natura*, An Essay on the First Book of T. Lucretius Carus (London, 1656), p. 3:

> Lucretius with a stork-like fate
> Born and translated in a State
> Comes to proclaim in English verse
> No monarch rules the Universe;
> But chance and Atomes make *this All*
> In order Democratical,
> Where Bodies freely run their course,
> Without design, or Fate, or Force.

Waller's use of "Democratical" (pertaining to democracy) as a play on "Democritical" (pertaining to the Greek atomist Democritus) is common to the period.

I believe, this description of chaos in which Milton allows himself to come closest to his most formidable English adversary in contemporary moral and political philosophy. By representing the abyss as the Hobbesian state of nature, Milton seems almost to reproduce Hobbes's unassuageable fear of political conflict, those endless skirmishes and disagreements that threaten, in *Leviathan*, to "Reduce all Order, Government, and Society, to the first Chaos of Violence, and Civill warre."[53] The random, purposeless violence of Milton's chaos seems also to work, at least temporarily, to assert the same political logic affirmed in Hobbes. Milton's chaos, like Hobbes's, functions to demonstrate the importance of the assertion of a sovereign authority. But the nature of this sovereign intervention differs tremendously for the two writers, and it is through his remarkable representation of Creation that Milton is able most fully to correct Hobbes's political premise. Whereas the chaotic life of man in the Hobbesian state of nature is redeemed through the direct and ongoing intervention of a powerful sovereign, the Miltonic chaos is transformed by Creation into an autonomous universe through a single, nonrepeatable act of divine infusion. When Raphael describes God's infusion of vital virtue and vital warmth throughout the fluid mass of the abyss, we see an image of Milton's own attempt to subsume, into his poem of radical Christian liberty, the very foundations of Hobbes's devastatingly authoritarian political philosophy.

The means by which this ideological subsumption occurs is another of the poem's physiologically suggestive natural processes. The Holy Spirit, Milton's name primarily for the energic principle behind most natural processes, inseminates and then, quite startlingly, digests chaos:

> Darkness profound
> Cover'd th' Abyss: but on the wat'ry calm
> His brooding wings the Spirit of God outspread,
> And vital virtue infus'd, and vital warmth
> Throughout the fluid Mass, but downward purg'd
> The black tartareous cold Infernal dregs
> Adverse to life; then founded, then conglob'd
> Like things to like, the rest to several place
> Disparted, and between spun out the Air.
> (7.233–41)

Elaborating on the processes of conglobing and disparting that constituted the self-determining motion of elements depicted by Uriel, Raphael, as I

---

53. Thomas Hobbes, *Leviathan*, ed. C. B. Macpherson (Harmondsworth: Penguin, 1951), p. 469.

have noted above, echoes the narrator's own description of the Holy Spirit's stunningly ambisexual behavior in brooding over and then inseminating the fluid mass of abyss. He narrates for us the moment of origin for the liberal, self-ordering cosmos, that extraordinary point in time when the form-less mass of elements is actually supplied with its vital power of self-determination.

But Raphael also describes a moment in this initial stage of Creation that cannot, I think, be reconciled with any of the natural philosophy presented elsewhere in the poem or by Milton himself in the *De doctrina*. The specific action described here, after insemination, is perhaps the most troubling nat-ural philosophical event in the poem, and this passage, I believe, holds the key to the puzzling political science of *Paradise Lost*. Before chaos could be organized into creation, Raphael explains, the Holy Spirit "downward purg'd / The black tartareous cold Infernal dregs / Adverse to life." The terms Milton employs for that which is excreted in the purgative step of this pro-cess—tartar and dregs—are the standard names in the period's natural phi-losophy for the inassimilable elements purged from the system in the process of digestion. In the *Treatise of the Rickets*, to cite just one example of the contemporary physiology of tartar, Glisson discusses the "tartar-like matter excreted in and with the Urin."[54] William Kerrigan has accounted movingly for the importance of the digestive process for Milton the man: the poor digestion and the violent flatulence of which Milton complained not only wracked his bowels with pain but, he believed, permitted unescaped vapors to rise into his head and to occlude, tragically, his eyesight. And Kerrigan, Michael Lieb, and Stephen Fallon have described how nearly every impor-tant event in *Paradise Lost* is marked by its figuration as a process of diges-tion.[55] Not only is the Creation figured in an explicitly digestive vocabulary, but the process whereby Adam and Eve might have metabolized themselves into angels (5.491–500) and the final separation of the saved from the damned at the Last Judgment (12.546–49) are imagined as digestive actions of salutary assimilation and purgation.

However central Miltonic digestion is to the figural world of the poem, we are still entitled to ask why it is that Milton's Creation requires this act of digestive purgation at all. We might reasonably have expected that the undigested lump of futurity Milton depicted as chaos in Book Two would be entirely transmuted at the Creation into the vital and animate body de-

54. Glisson, *Treatise of the Rickets*, p. 177.

55. See Kerrigan, *Sacred Complex*, pp. 193–245; Michael Lieb, *The Dialectics of Creation: Patterns of Birth and Regeneration in "Paradise Lost"* (Amherst: University of Massachusetts Press, 1970), pp. 16–34; Lieb, "Further Thoughts on Satan's Journey through Chaos," *Milton Quarterly* 12 (1978): 126–33; and Fallon, *Milton among the Philosophers*, pp. 102–6, 209–10.

scribed in *Christian Doctrine* or in Uriel's monistic creation account in Book
Three. When the Spirit of God hovers over the vast abyss and makes it
pregnant, chaos is infused with a vital spirit that permits it to generate a
world with relative autonomy. But here, at this central moment in Raphael's
creation narrative, the Son's insemination of the vital power of self-motion
is revealed not to have fully transformed the matter of chaos from a Hobbesian world of inert, colliding bodies to a Glissonian world of universal animation. In one remarkable line, Milton appears to indicate that there
remained a portion of the deep for which the process of spiritualization
simply did not take: "The black tartareous cold Infernal dregs" refuse assimilation into the vitalist world of Milton's poem.

Like the warring atoms in the chaos of Book Two, the recalcitrant creational residue here demands interpretation. How do we account for the intractable presence in the Miltonic universe of these material dregs "Adverse
to life," when the *Christian Doctrine*, and the general logic of Milton's monistic science, insists that some type of virtue or divine spirit had been infused into the whole of the original matter? A truly monist universe,
permeated entirely with divinity, cannot logically accommodate a pocket of
inertness that has escaped divinization. Just as the noun *dregs* announces a
noisy thematic disruption through its modification by no fewer than four
contiguous adjectives, the appearance in Milton's science of this lifeless, spiritless waste violently interrupts the gradualist monist continuum of matter
and spirit. The tartareous dregs of creation introduce into the otherwise
monistic world of the poem a residual trace of dualism.[56]

A preliminary explanation of the strange presence of inanimate matter in
Milton's animate universe can be drawn from the Renaissance science of
digestion. Raphael's image of this moment of purgation is inscribed, I believe, in a lively scientific debate about the indigestible elements of food
known as dregs, or tartar. Milton's ultimate source for this digestive lore is
the early-sixteenth-century alchemical philosopher Paracelsus, who posi-
tioned the science of tartareous dregs at the heart of an entire philosophy
of natural and divine justice.[57] Using as his figural base the tartar, or calcified
wine, that deposits itself along the lining of wine casks, Paracelsus struggled

---

56. That this moment of infernal purgation is not to be confused with the prior moment of
the creation of Hell, a Christian version of the classical Tartarus, is made clear by Raphael's
careful time-scheme (7.132–38). See Alastair Fowler's chronology of the poem's events in *John
Milton: Paradise Lost* (Essex: Longman, 1971), pp. 26–27. The theological confusion invited by
this passage in Book Seven has been observed by David Aers, in *Paradise Lost: Books VII–VIII*,
ed. David Aers and Mary Ann Radzinowicz (Cambridge: Cambridge University Press, 1974), p.
27; and by Kendrick, *Milton*, p. 204.

57. For a discussion of the Paracelsian philosophy of tartar, see Pagel, *Paracelsus: An Introduction*, pp. 153–65.

to make sense of the human pathology he called *"Duelech,"* the hardened residue that forms stones within the bladder and kidneys and that seemed to him the same hardened substance with the "cadaverous smell" that we still call tartar of the teeth. As one of the only types of matter "incapable of receiving the breath of life," resisting infusion by the vital warmth and virtue of divinity, Paracelsian tartar resides naturally in water and vegetation and is drawn into the body through the unavoidable ingestion of food.[58] Most significantly for Paracelsus, the intractable dregs of matter known as tartar are the chief source of misery in the lives of men, "the most inward immediate causes of all Diseases, the daily Nurses of the calamity of mortals."[59] Given the scope of the devastation wreaked by tartar, Paracelsus was prompted to ask the question no doubt still asked today: Where does this stuff come from? His answer was essentially theological. More consequential even than thistles and thorns, tartar is nothing other than the chief manifestation of evil generated at the Fall.[60]

Milton may have found in Paracelsus a precedent for the location of a stubbornly inanimate waste within an otherwise animate body of matter. But the example of Paracelsus certainly provides no logical justification for the hypothesis of unspiritualized dregs in the seamlessly spiritualized matter *before* the Fall. Paracelsus himself would have considered absurd the presence of tartareous dregs in an unfallen abyss.[61] It is in fact the problem of the inescapable fallenness of dregs and tartar that led one of Milton's vitalist contemporaries to mount a full-scale attack on the Paracelsian theory of tartar. Jean Baptiste van Helmont's critique of Paracelsus, collected for publication by his son in 1648, exposes with an extravagant argumentative ingenuity the injustice of any theory that locates the origin of tartareous dregs outside man himself. I propose that we turn now to van Helmont, surely one of Milton's chief sources for his theodicial philosophy of fermentive vitalism, because the Helmontian reading of the tartar hypothesis proves invaluable as we untangle the troubling implications of the strange act of tartar control at the heart of Milton's creation.

58. Paracelsus is quoted here from Milton's likely source for this thesis, van Helmont's 1648 *Ortus Medicinae*, cited here in the English translation, *Oriatrike*, p. 246.

59. Paracelsus, quoted in van Helmont, *Oriatrike*, p. 231.

60. The early modern physiological identification of tartarous dregs with primordial evil is intuited by Kendrick, who remarks provocatively in *Milton* that the "principal romance embodiment of evil is surely that of the 'dark infernal dregs,' the slime and ooze purged by the Son from the original matter" (p. 204).

61. Paracelsus denied the existence of any creational residue: "A woodcarver is able to carve out of wood whatever he pleases provided he can separate the wood from that which does not belong to it. Thus also God took out, drew out, and separated all His creatures from one mass and material, and He left no chips in the process" (Paracelsus, *Selected Writings*, p. 14).

It is a commitment to the justification of the ways of God that motivates
the Helmontian consideration of tartar. If, as Paracelsus had suggested, the
tartar that deposits itself on teeth or in kidneys actually derived from natural
water and vegetation, then, according to van Helmont, we are obliged to
impugn God and not man for the punishment of death: "Seeing no tartar is
read in Genesis to be created forthwith after sin . . . therefore the Creator
[would have had to have made] the punishment before the fault." Sensitive
to the fact that any such "punishment before the fault" can raise only the
gravest questions about divine justice, attesting to the unjust, even evil, foundations of the entire creation, van Helmont took it upon himself to defend
God from the Paracelsian thesis:

> I do not admit of a tartarous generation . . . appointed by God for our destruc-
> tion. . . . For death is not the handy-work of God: And God saw that whatsoever
> things he had made, they were good; as well in his own intention of goodness,
> as in the essence of the Creature: Therefore there is no matter in the waters,
> which was created to stir up the Tragedy of Tartar, or a *Duelech* in us. More-
> over, if there be any evil now, or that may come to passe among the digestions,
> surely that is not from the Creation, appointment, property, efficient of matter,
> and the finall intention of the Creator; but doth issue wholly from our errour,
> and the corruption of nature.[62]

Like Milton's angel Michael, who names the "Intestine Stone" as one of the
consequences of Adam's sin (11.484), van Helmont attributes the origin of
tartar, or *Duelech*, to "our errour, and the corruption of nature" after the
Fall: if a tartarous tooth "do thrust forth" a "stinking superfluity . . . that
excrement is not so much the superfluity of meats, as the excrement of
man."[63] He denies on the grounds of theodicy any "tartarous generation . . .
appointed by God." If there is a "matter in the waters, which was created
to stir up the Tragedy of Tartar," then we would have to ascribe evil to the
"appointment, property, efficient of matter, and the finall intention of the
Creator," an ascription that can only point, absurdly, to a hidden Manichean-
ism at the Creation's beginnings. If we extend the logic of van Helmont's
analysis of Paracelsus to our own reading of Milton's Creation, we can come
closer to the terrible implications of Raphael's inclusion of the "black tartareous cold Infernal dregs" as an externally generated, precreative substance
that must be purged from the original matter. It is not merely that the
tartarous dregs have interrupted the monistic continuum of variously graded
strata of spiritualized matter. The presence of these dregs empties Milton's

62. Van Helmont, *Oriatrike*, p. 250.
63. Van Helmont, *Oriatrike*, p. 247.

natural philosophy of its theodicial force. The radical moral power of Milton's animist materialism quite simply evaporates if even a portion of the material universe cannot be shown to have derived from the intrinsically good substance of an intrinsically good God. The tragedy of tartar can only draw in its inevitable logical wake the tragedy of the Fall.

In three simple lines, Milton, it can safely be said, has sabotaged his attempt to justify the ways of God to men. Before, however, we dismiss the poem's seemingly reckless resignation to an intractable theological aporia, we might do well to consider A. S. P. Woodhouse's observation on the theological incoherence of the chaos Milton shows us in Book Two: "It would appear that this is a relic of some anxious consideration of the problem of evil."[64] The presence in the Miltonic Creation of any evil at all is naturally a cause of concern for students of Milton's theology, which is determined to drain the positivity from evil by defining it as a simple absence of good. For Regina Schwartz, the apparent evil of chaos suggests the poet's attempt to make a theological point; Milton uses the Manichean figure of God's "primordial battle with chaos" to render all the more glorious the eventual subsumption of chaos into the greater good of divinity.[65] But as Schwartz herself notes, an "evil," or at least excessively disordered, chaos cannot be accommodated so comfortably within the confines of Milton's rigorously monistic, *ex deo* theology. Woodhouse is right, I think, to suggest that the initial representation of chaos in *Paradise Lost* is "a relic of some anxious consideration." But we can ascribe this anxiety to a source more complex than the strictly theological question of the origin of evil. Given what was in the 1650s and 1660s the cultural pressure inevitably bearing down on any representation of material organization, we can see how the "evil" dregs of chaos function significantly in the poem's politicized speculation about the nature of the original matter. The theologically irreconcilable figure of the "dregs" constitutes a relic of those political anxieties that trouble Milton's final political writings, especially *The Readie and Easie Way to Establish a True Com-*

---

64. "Notes on Milton's Views on the Creation: The Initial Phases," *Philological Quarterly* 27 (1949): 229n. Woodhouse cites a Valentinian precedent for this disturbing purgative moment in Milton's creation, one like the Paracelsian text that locates a preexistent evil in the dregs of the original matter: God "began to work upon [chaos] and wished to separate its best parts from its worst, and thus made all that was fitting for God to make out of it. But so much of it was like lees, so to speak, this being unfitted for being made into anything, he left as it was . . . , and from this . . . what is evil has now streamed down among men" (Methodius, *Concerning Free-Will*, trans. Clark, *Ante-Nicene Fathers* [Buffalo, 1880], 6:358; quoted in Woodhouse, pp. 230–31n). Chambers argues for the evil of Milton's chaos, in "Chaos in *Paradise Lost.*"

65. *Remembering and Repeating*, p. 32. Schwartz claims earlier that Milton's "is a chaos that is hostile, an 'Other,' but one which God has created. Ultimately, subsuming this Other, evil, to divinity is the project of his theodicy" (p. 17).

monwealth (1659), a treatise everywhere concerned with the unstinting encroachment of political chaos.

We have explored the connection between the glimmers of vitalism in *Christian Doctrine* and *Paradise Lost* and the political idealism Milton voiced at the height of the Vitalist Moment in his *Defence of the English People*. We must turn now to Milton's last major political treatise to grasp the full significance of the figure of the unanimated dregs that works to undo the necessarily universal spiritualization of matter. There are moments, in *The Readie and Easie Way*, just as in the poem, in which Milton re-engages the most radical aspects of his natural and political philosophy forged at the Vitalist Moment. In an uncharacteristically liberal proposal in *The Readie and Easie Way*, Milton suggests that the civil administration of the entire Commonwealth be decentralized by pressing local "nobilitie and chief gentry" to "bear part in the government, [and to] make thir own judicial laws" (6:144), an innovative argument for local self-government that, according to one historian, positively "contributed to democratic theory" in the West.[66] These local leaders, Milton writes excitedly, could even establish "schools and academies at thir own choice" (6:145). When Milton imagines the possibility of this deregulated dissemination of education and culture—the sociopolitical analogue of Raphael's "vital virtue"—he conjures an image drawn directly from the radical physiology of a monist such as Francis Glisson. This program of local educational development, Milton writes, "would soon spread much more knowledge and civilitie, yea religion through all parts of the land, by communicating the natural heat of government and culture more distributively to all extreme parts, which now lie numm and neglected" (6: 145). In this brief entertainment of the popular sovereignty he had anticipated in 1649 and 1651, Milton returns to the natural philosophy of the Vitalist Moment, the belief in the equal diffusion of soul, or vital warmth, throughout the entire human body and throughout the entire creation. The pathology of rickets, for Glisson, was the result of the unequal distribution of natural heat: the center of the body may be hot, but the "External parts are . . . affected with a cold and moist Distemper, and a benumedness of the natural Spirit."[67] The civil numbness of which Milton complains is amenable to the same solution as the Glissonian body. Given an equitable dispersal of the state's virtuous resources, "the natural heat of government and culture," the rachitic members would be strengthened, the body politic made once again whole.

---

66. Don M. Wolfe, *Milton in the Puritan Revolution* (New York: Humanities Press, 1963), p. 303.

67. Glisson, *Treatise of the Rickets*, p. 34.

Despite this fond glimpse into a localist and vitalist future, Milton is in this treatise far more overwhelmingly struck by the seeming irredeemability of his conservative countrymen; he finds at the present moment, in 1659, a multitude of individuals so unregenerate and so unremittingly royalist that he can only maintain for a brief paragraph this vitalist hope in local self-government. The later versions of Miltonic philosophy—both natural and political—point therefore not toward a utopian egalitarianism but toward a hierarchical segregation of elements according to what is essentially their degree of spiritualization. In *The Readie and Easie Way*, written just weeks before the Stuart Restoration in 1660, Milton proposes the empowerment of a "Grand Councel" of regenerate saints whose authority over the masses derives from their self-sufficient powers of reason. Although these hundred or so good men enjoy an equality among themselves as they serve on the council, governing equally according to the laws of reason written on the fleshly tables of their hearts, the liberal organization of the council leaders exists nonetheless as a circumscribed zone of autonomy within an otherwise hierarchical system.

When it is incumbent on Milton to account for the mechanics of a political system that empowers a spiritually minded elite to impose order on a recalcitrant majority, it is once again an image from natural philosophy that he employs. Milton tackles the difficult question of the selection of the Grand Council in a society unable to turn for its important decisions either to a centralized sovereign or to the irrational multitude. How is the Grand Council to be chosen? Milton proposes a stunningly complex election process intended to institute a form of political consensus not achievable in an unruly popular government:

> Another way will be to wel-qualifie and refine elections: not committing all to the noise and shouting of a rude multitude, but permitting only those of them who are rightly qualifi'd, to nominat as many as they will; and out of that number others of a better breeding, to chuse a less number more judiciously, till after a third and fourth sifting and refining of exactest choice, they only be left chosen who are the due number, and seem by most voices the worthiest. (6:134)[68]

The science informing this passage is not of course the vitalist science of material self-motion; it is a mechanistic description of an externally motivated "sifting" of inert particles that can be forced into separation by their quan-

---

68. For a different perspective on the natural philosophy of Milton's description here, see Fallon, *Milton among the Philosophers*, pp. 109–10.

titative size and weight. This electoral process of sifting and refining institutes on the political level the analogous disposition of unequal elements that we find in Raphael's digestive account of creation: the least worthy elements are eliminated, and the worthiest and most spiritual remain to establish the new order. The beauty of Milton's description of this awkward, four-step election process is crystallized in the passive verb employed in the last sentence of the passage: "they only be left chosen who are the due number, and seem by most voices the worthiest." The passive voice is crucial here, because it allows Milton to remove from this political process any hint of authoritarian agency, at the same time that it identifies this electoral institution of consensus-identification as a public form of rational choice.

Don M. Wolfe has located Milton's model for this electoral scheme in Plato's *Laws*, and it is instructive to look briefly at the revision to which Milton subjects his Platonic original.[69] For Plato, according to Wolfe, "the franchise should be limited to those having a record of military service. These men nominate as many as they wish by engraving their names on tablets and bringing them to a central place of voting. The three hundred nominees whose names appear most often on the tablet are then again voted on by all soldiers, each elector bringing a tablet inscribed with the name of his favorite nominee among the three hundred."[70] The hundred nominees with the most votes in this second-round election are then reduced a third time, the final selection of the thirty-seven magistrates being made, like the first selection, by the entire voting population. Milton borrows from Plato the institution of a series of electoral rounds, adding for good measure a fourth and final "refining of exactest choice." But he has also clearly extended the delimiting effects of this electoral process of sifting and refining. Milton proposes to winnow out not only unworthy magistrates but unworthy voters as well: in Milton's revision of Plato's scheme, only "those of them who are rightly qualifi'd" are permitted to vote in the first round, and the successive rounds of elections have even smaller pools of voters, "each of a 'better breeding' than the one preceding."[71] Milton, perhaps wisely, omits from his treatise an explanation of how the political determination of "better breeding" is to be made. But the overwhelming impracticality of the entire proposal suggests

---

69. Wolfe, *Milton in the Puritan Revolution*, pp. 300–302. See Plato's *Laws*, trans. R. G. Bury (New York: Putnam, 1926), 1:401.

70. Wolfe, *Milton in the Puritan Revolution*, p. 302.

71. Ibid. There may have been some justice to the pamphleteer's claim in *Censure of the Rota* that Milton's election scheme, and his corollary proposal of a life term for those members elected, would lead ironically to a government as "Arbitrary and Tyrannical" as the one it replaced: "For though you bragge much of the Peoples Manageing their own affaires, you allow them no more share of that in your *Utopia* (as you have ordered it) then only to set up their throates and Baul . . . once in an Age" (pp. 16, 14).

the lengths to which he is willing to go to protect the Commonwealth from the "noise and shouting of a rude multitude." England's Parliament had of course undergone a succession of purges since 1648. But the electoral purgation Milton is counseling now is of an entirely different order. It is not just that the dregs must be eliminated from the Grand Council; they must be eliminated as well from the body of the electorate.

Milton may have been sincere in *The Tenure of Kings and Magistrates* and *The Defence of the English People* in affirming the possibility of a political system in which sovereignty could be extended and distributed equally among the rational, spiritually minded citizens of England. But his own experience with the republican experiment failed to bear out the viability of his utopian ideals. By 1659, a majority of English citizens were responding to the conditions of near political chaos by voicing their desire for the return of the Stuart monarchy. This "misguided and abus'd multitude," as Milton called this majority at the end of *The Readie and Easie Way*, were not only "chusing them a captain back for *Egypt*" (6:149). They were also rendering it impossible for Milton to maintain a faith in his vitalist principles and their implications for both political and natural philosophy. God may very well not have extended and diffused his spirit throughout the mass, or the masses, of his creation. Milton's enemy Salmasius, it seemed now, might have been right: England was inhabited primarily not by rational and like-minded men, as Milton had insisted in the *Defence*, but by those individuals Salmasius had dismissed as the "dregs of the common people."[72]

I suggest that it is for the sake of enlisting his representation of chaos into the contemporary conversation about the ontology of political societies that Milton is compelled to present a natural philosophy so at odds with his best theological impulses in the *Christian Doctrine*. The contradiction between the chaoses represented in the theological treatise and in the epic is the symptom of the inescapable pressure of political philosophy on Milton's theological science. The discrepancy between a vital chaos and a chaos that harbors a corner of inertness measures the distance between the faith Milton held in popular sovereignty in the period between 1649 and 1651 and his soberer resignation, articulated most fully at the end of the republican decade, to a perpetual parliament of like-minded oligarchs empowered to exercise force over a multitude that showed no signs of capacity for the popular sovereignty Milton had so eagerly anticipated. The irrational, unspiritualized

72. In the *Defense of the English People*, Milton refers twice to Salmasius's scornful reference, in the *Defensio Regia*, to the "dregs" (*colluvies*) of the English commoners (7:34, 7:38). Another Salmasian term for the English, usually translated as "dregs," is *faeces* (7:68), meaning "tartar" or "sediment."

dregs had established themselves as permanent and untransformable members of the body politic.

The infusion of divine virtue into the reasonless anarchy of chaos, I have argued, bespeaks Milton's attempt to confront, and overgo, the authoritarian conclusions Hobbes drew from his similar identification of the masses as unruly and irrational. But it must also be confessed that Milton's alignment of the debased Hobbesian vision of the state of nature with his own uncreated chaos invites a darker, less optimistic set of associations. Satan, in Book Two, promises Chaos that he will work to return to its original chaotic state the belated imposition of creation, the animate and orderly divine matter that had been "Won from the void and formless infinite" (3.12). Satan's boastful promise to Chaos is not, of course, to be trusted, and the poem gives us no powerful theological reason to fear this reduction of creation to "her original darkness" (2.981–87). But Satan's threat, however unsound theologically, resurfaces later in the poem, as Adam remarks that the stars shine in the sky "Lest total darkness should by Night regain / Her old possession" (4.665–66). I suggest that this ontologically absurd threat, repeated twice in the poem, can be found to make sense only once this natural philosophical speculation is translated into an organizational discourse whose significance is political. The possibility of a chaotic resurgence has no meaningful role in the poem's cosmology, but its expression voices Milton's fear, perhaps not so unsound, of an ever-encroaching political chaos.

The Readie and Easie Way is charged with the pathos of the anarchic conditions of the republic in the last years of the decade. Resorting to an emotional deployment of the imagery of chaos, Milton decries the return of the "old encroachments" on England's hard-won liberties that would certainly destroy the newly created Commonwealth. A vision of the "approaching ruine" brought on by what Milton calls the "distracted anarchy" of the people clamoring for a monarch is invoked throughout the tract as a call to the English to return to their vital capacity for self-government. But this call, Milton knows, will not be heeded in the unreasoning drift toward monarchy, "which the inconsiderate multitude are now so madd upon" (6:134).

Although Milton may not wish us to credit Satan's promise for the return of all created matter to the originally anarchic state of chaos, his wearying focus on the possibility may express a doubt that has its historical (and biographical) origins in the political defeat of the republican Commonwealth and the failure of the much-anticipated revolution of the saints. The threat in *Paradise Lost* of an ultimate resurgence of chaos articulates Milton's doubt about the stability and the ultimate efficacy of the goodness and virtue presumably so intrinsic to the organization of the created universe. The digestive process of distillation that works continually to remove from organism or organization its tartareous dregs may well prove insufficient to ward off a resurgence and eventual usurpation by those elements of the organization "Adverse to life." It is the constant pressure chaos is imagined to apply on creation that signals, perhaps more than anything else in Milton's epic, that the Vitalist Moment at midcentury was never to be recovered. That ebullient intellectual ferment, at once political and scientific, could now be viewed solely through the lens of atavism and spoken of as the Good Old Cause.

5 Milton and the Mysterious Terms of History

> But when mankinde once sees, that his teacher and ruler is within him; then what need is there of a teacher and ruler without; they will easily cast off their burden.
>
> —GERRARD WINSTANLEY, *Fire in the Bush* (1650)

As it threads through the wandering mazes of free will and providence, *Paradise Lost* broods continually on the problem of agency, both human and divine. In establishing the nearly autonomous origins of Creation in chaos, Milton opened his poem to the political resonances of the dialectical relation of an often self-governing matter to the imposing power of divine will. But the epic's shift in emphasis after Book Eight from the natural philosophical question of creation to the more explicitly theological question of sin and punishment inhibits the daring poetics of vitalist agency that Milton explored earlier in the poem. In the depiction of the origins and consequences of the Fall, the later books of the epic seem to lack the glimpses into the radical organizational politics of the Vitalist Moment with which Milton charged the accounts of Creation. The deliberate vagueness concerning the politically urgent problem of the agents of change, and of punishment, is nowhere more self-consciously expressed than in the portrayal of the Father's determination to exercise justice on the figure most instrumental to the Fall: "yet God at last / To Satan first in sin his doom appli'd, / Though in mysterious terms, judg'd as then best" (10.171–73). The terms of the transformation Satan will undergo, in his spectacular Ovidian metamorphosis from fallen angel to serpent and his projected defeat by the Son, are unquestionably mysterious: we are left merely to speculate about the mechanics of God's application of Satan's doom, about God's causal role in this strange alteration, and about the relation of this curiously ad hoc execution of poetic justice, "judg'd as then best," to the workings of "Eternal Providence." Like all the changes after the Fall in *Paradise Lost*, Satan's punishment is represented explicitly as an act of justice; his doom constitutes a single instance in the judicial process that is always coincidental in Milton's justifiable universe with the course of history itself. In the case of the "mysterious terms" that govern Satan's eventual punishment, Milton slyly withholds any detailed justification of the ways of God. But in the remaining two books of *Paradise Lost*, he resumes the problem of the terms of providential agency and cosmic justice. With an almost disquieting degree of precision and detail, Milton delineates in these last books an authoritarian cosmos the human element of which comes to be structured by a dispiriting pattern of human sin and divine punishment. The agential and organizational implications of this pattern are the focus of my examination here of the epic's late authoritarian turn.

It is perhaps because Milton's epic can seem so complex an etiology of original sin, so intent on the subtlest matters of motive, the slightest precipitants to alteration, as well as on the tragic effects of history's heaviest change, that the last two books, which address directly the subject of human history, strike most readers as a disappointment. When in Books Eleven and Twelve Milton justifies the ways of God as they have been revealed by Scripture, he does so by staging a lesson in future history that has traditionally been seen to lack the expansive, humanistic sense of self-motivated agency the rest of the poem so strongly evinces. But since the late 1960s, there has been a welter of criticism defending the historical outlook that shapes the bleak prophecies Michael vouchsafes to Adam. Armed with the critical intent of apology, Miltonists have embarked on the "recognized and official project" Stanley Fish has identified as "the rehabilitation of Books XI and XII."[1] Careful to position the prominent seventeenth-century poet and statesman more or less comfortably within identifiable schools of historiographical discourse, scholars have studiously unveiled those structures internal to the last books that are seen to illuminate Milton's historical consciousness.[2] But in

1. Stanley Fish, "Transmuting the Lump: *Paradise Lost*, 1942–1979," in his *Doing What Comes Naturally: Change, Rhetoric, and the Practice of Theory in Legal and Literary Studies* (Durham: Duke University Press, 1989), p. 275.

2. The many interpretations of the conclusion to Milton's epic have tended to reproduce the theoretical problematic that impelled the disciplines of theology and history in Milton's time— the uneasy relationship between supposedly free human action and divine control. It has been variously argued that the last books maintain Milton's youthful, revolutionary faith in a divinely ordained historical progress; that the poet sees providence as guiding history in a direction that is inescapably cyclical and degenerative; and that he believes that history can be progressive only through some form of disciplined cooperation between God and man. Dennis H. Burden, in *The Logical Epic: A Study of the Argument of "Paradise Lost"* (London: Routledge, 1967), p. 182, and Hill, in *Milton and the English Revolution*, pp. 388–90, claim that Michael's lesson ultimately reinforces the value of human action and worldly progress through Christian heroism. The poem confirms the notion of "Satanic necessity" and "strict determinism" for J. B. Savage, "Freedom and Necessity in *Paradise Lost*," *ELH* 44 (1977): 307. Those who focus on Michael's concluding summary to find that God has willed a historical degeneration include Joseph H. Summers, "The Final Vision," in *Milton: A Collection of Critical Essays*, ed. Louis Martz (Englewood Cliffs, N.J.: Prentice-Hall, 1966), p. 203; Patrides in *Grand Design of God*; Robert L.

their attempts to justify Michael's particular version of the ways of God, these critics have displaced the eccentric theology of the last books onto the center of the poem, forging an identification between Michael and Milton that confuses the larger function of Michael's disconcertingly providential history within the poem as a whole.[3]

I, too, am interested in exploring the mysterious terms that constitute Milton's elusive philosophy of historical agency, both human and divine. But I do not want to limit this understanding to an analysis of the recognizably "historical" discourses that surface in Books Eleven and Twelve. Nor do I want to dismiss as an invalid interpretive key the sense of disappointment that so often accompanies a reading of those books. Throughout *Paradise Lost*, as we saw in Chapter 4, implicit figurations of historical change emerge in contexts outside the conventionally historiographical exposition of world events presented by Michael. "Say first what cause," Milton bids the Heavenly Muse near the beginning of the poem, invoking the epic tradition to introduce the problem perhaps most central to the discipline of history, the determination of cause. In justifying the causes and effects of a single human action, the original sin, Milton engages the discourses of theology, physiology, and both moral and natural philosophy. I explore here some of these less obviously "historical" considerations of cause and effect that shape the end of Milton's narrative, looking in particular at the epic's climactic scene of expulsion. The various accounts of the expulsion, I believe, force on the poem's conclusion a conflict among competing philosophies of agency that bespeaks the larger contradiction in the Miltonic discourse of cosmic and therefore political organization.

As illustrated by his Ramist treatise, *The Art of Logic*, and his rigorous use of Ramist method in his *Christian Doctrine*, Milton is often led to ground his analysis of the consequences of human action in the logical determination of cause and effect. His Protestant temperament no less than his engagement of contemporary science led him to question the validity of sacramental or symbolic connections among events, and it is therefore as a series of logical consequences of that pernicious cause, the Fall, that he presents the course

---

Entzminger, "Michael's Options and Milton's Poetry: *Paradise Lost* XI and XII," *English Literary Renaissance* 8 (1978): 206; and Georgia B. Christopher, *Milton and the Science of the Saints* (Princeton: Princeton University Press, 1982), p. 189. Mary Ann Radzinowicz argues for Milton's depiction of the interaction of human and divine efforts, in *Toward "Samson Agonistes": The Growth of Milton's Mind* (Princeton: Princeton University Press, 1978), p. 76. See also Radzinowicz, "Man as a Probationer of Immortality: *Paradise Lost* XI–XII," in *Approaches to "Paradise Lost": The York Tercentenary Lectures*, ed. C. A. Patrides (Toronto: University of Toronto Press, 1968), pp. 31–51.

3. See the discussion of the prevailing misreading of Milton's theology in Rumrich, "Uninventing Milton."

of human history in *Paradise Lost*. Milton, of course, would never have embraced a materialism like that of Hobbes or Descartes which excluded spirit as an integral component of the physical universe; as a theologian with ties to the vitalist science of van Helmont, Harvey, and Glisson, Milton asserted a monistic materialism that posited the infusion of body with spirit. He did, however, share with all his scientifically minded contemporaries the growing interest in the intermediate processes of natural motion. Although his work everywhere acknowledges the primacy of first and final causes, his *Christian Doctrine* and *Paradise Lost* convey a particular emphasis on the philosophical and ethical implications of the new material laws of motion; they comprise a unique version of the late-seventeenth-century project of reconciling the mechanistic conception of nature with the Christian idea of providence. I argue in this examination of the last two books of *Paradise Lost* that his decidedly idiosyncratic science of vitalism supplies Milton with a compelling means of asserting, at the poem's end, a revolutionary discourse of agency and organization. In the face of Michael's overwhelmingly authoritarian theology of divine punishment, Milton surreptitiously revives his vitalist critique, at once egalitarian and secular, of the centralized form of political organization to which he referred throughout his political career as "tyranny."

The Law God Gave to Nature: Milton's Historical Materialism

As the epic's title suggests, the point at which the first parents actually lose their paradise constitutes an important narrative crux. I propose that we assess some of the political and natural philosophical issues informing the last books by examining two passages that articulate the actual loss of Paradise: the Father's explanation to his Son of the necessity of the expulsion and his subsequent induction of his warrior angel Michael to effect the first parents' leave-taking. In the first scene, the Son has been pleading with the Father that man's sins be engrafted onto the Son himself. In a display of mercy, the Father accepts the Son's sacrifice and promises that man will indeed be saved; but the Father then qualifies his offer of redemption with the following condition:

> But longer in that Paradise to dwell,
> The Law I gave to Nature him forbids:
> Those pure immortal Elements that know
> No gross, no unharmonious mixture foul,
> Eject him tainted now, and purge him off
> As a distemper, gross to air as gross,

And mortal food, as may dispose him best
For dissolution wrought by Sin, that first
Distemper'd all things, and of incorrupt
Corrupted.

(11.48–57)

In explaining to the Son the limitations of his mercy, the Father describes
what will be the necessary expulsion of Adam and Eve from the garden. In
the context of the Son's request for mercy, the Father's account of this
particular punishment of Adam and Eve assumes the status of theodicy:
given the considerable investment the poem has generated in the notion of
paradise, the theodicial project behind *Paradise Lost* requires at least an
attempt to absolve God of direct responsibility for its loss. Although, the
Father leads us to understand, he would like to pardon fully the sinful pair,
the laws of nature established at the Creation prohibit man's further habi-
tation in the garden. Adam had expressed in Book Ten some doubt that God
could "extend / His Sentence beyond dust and Nature's Law" (10.804–5).
Confirming what may have seemed at the time a rather far-flung speculation
by a fallen creature, the Father himself cites the circumscription of his power
by "Nature's Law." He justifies what might otherwise appear to be a callous
lack of mercy by an appeal to an even higher principle of irrevocable natural
law. Milton's contemporary, Robert Boyle, had in 1666 attempted to head
off any such suggestion that God's personal authority might be subject to
material forces beyond his control: God "is not over-ruled, as men are fain
to say of erring nature, by the head-strong motions of . . . matter, but
sometimes purposely overrules the regular ones to execute his justice."[4] Milton's Father, though, pointedly disavows in his discussion of justice the
crutch of personal rule supporting the authoritarian God of Boyle. In a vitalist, and therefore antiauthoritarian, description of matter in motion, the
Father represents the heavy change of the expulsion as a lamentable reaction
of the material body of Paradise to the newly corrupted bodies of Adam and
Eve. "Earth, Flood, Air, Fire," those "pure immortal Elements," cannot but
follow the irrevocable laws of physical and physiological equilibrium and the
motion of bodies (3.715, 11.51). These eternal operations of nature that
preserve the purity, and the liberal constitution, of Paradise will themselves
automatically respond to the Fall by ejecting the tainted human body from
the garden's harmonious air.

Although Milton's critics have traditionally dismissed this passage as the

---

4. Boyle, *Free Inquiry into the Vulgarly Received Notion of Nature*, in *Works of Boyle*, 5:
198.

Father's temporary indulgence in a fanciful literary conceit, the poem itself casts no doubt on the narrative validity of this description of the mechanics of expulsion. In fact, this monistic description of the expulsion simply extends into the human action of the poem the vitalist metaphysics of matter that, as we saw in Chapter 4, Raphael has already expounded to Adam (5.404–505). Just as the body is inseparable from the spirit in Milton's monistic ontology, the workings of the Father's harmoniously balanced, material universe are indissoluble from the moral laws of heaven: by reason of the vital spirit infused at the Creation into the original matter, material processes exercise their own highly moral form of justice as they follow inexorably the laws of nature the Father has established. Nature, according to Raphael, is "one first matter all," and this fully spiritualized material universe permits morally superior substance to "sublime" until "body up to spirit work" (5.468–90). The animism of the poem's universe thus allows matter to proceed up to the maker "If not deprav'd from good" (5.471). Attending the Fall of man, however, is a certain "dissolution wrought by Sin," and like the rebel angels, who were "Purest at first, now gross by sinning grown" (6.661), Adam and Eve have undergone a process of physical deterioration, or, rather, a solidification of bodily matter that before the Fall had been lighter and more attenuated. It is precisely that natural, alchemical process by which Adam's and Eve's "bodies may at last turn all to spirit, / Improv'd by tract of time" (5.497–98), which must now react to the new disproportion of humors and elements brought on by the couple's ingestion of the forbidden fruit. Carefully maintaining the atmospheric equilibrium of its material organization, Paradise must now purge the gross bodies of the dissipated pair.[5]

Gerrard Winstanley had in 1649 described the distempering physical effect of sinful Adam on the material organization of Paradise: "the poyson of mans unrighteous body, dunging the Earth, filled the grasse and herbs with strong unsavory spirits, that flowed from him."[6] When Adam's dead body was later buried in the earth, the contagion of his lapsarian effluvia was so powerful it distempered permanently the elemental balance: "he corrupted the whole creation, fire, water, earth and aire, and still as the branches of

---

5. The material interaction between the human body and its atmospheric surroundings seems to have interested Milton as early as the 1640s. In the third draft of his proposed dramatic treatment of *Paradise Lost*, which survives in the Trinity Manuscript, Milton planned to begin the tragedy with Moses in Paradise, "recounting how he assumed his true body, that it corrupts not because of . . . the purity of the place that certain pure winds, dews, and clouds preserve it from corruption" (quoted in *Milton: Paradise Lost*, ed. Fowler, pp. 3–4). Kerrigan places Milton's image of Moses here in the context of Helmontian science, in "Heretical Milton," p. 140.

6. *Works of Winstanley*, p. 114.

his body went to the earth, the creation was more and more corrupted."[7] It is in an effort to prevent this unsavory Adamic pollution of the "pure immortal Elements" of Eden, described so vividly by Winstanley, that Milton has thoughtfully installed in his Paradise an expulsive mechanism that triggers automatically at the intrusive presence of a polluting agent. Milton may have found a scriptural precedent for this bodily act of expulsion in God's warning to the sinful Hebrews that "the land itself vomiteth out her inhabitants" (Lev. 18:24–28). But the figuration of the expulsion points, I believe, to a physiological process other than emesis, one indicated in Michael Hollington's provocative note that Paradise here seems to function as a "body expelling diseased tissue."[8] The physiological paradigm shaping Milton's image for this organismic expulsion of polluted matter could very well be the recently formulated principle of tissue irritability. Milton, I suggest, evokes here the vitalist principle of decentralized bodily response he would have found in Francis Glisson's De rachitude of 1650. Like the bodily tissue that for Glisson responds locally to irritants and toxins, the body of Milton's Paradise responds automatically, and without reference to a higher, central power, to purge itself of tainted flesh.

Despite its many affiliations with the monistic universe described by Raphael and the liberal science of the Vitalist Moment, the Father's naturalistic representation of the expulsion is most striking for its utter lack of resemblance to the elaborate narrative of expulsion we encounter in the remainder of Books Eleven and Twelve. The poem, of course, does not continue in this vitalist vein to represent the expulsion as the unavoidable consequence of autonomous material forces, as some fantastic struggle between warring humors and elements that ends, in a manner more suited perhaps to science fiction, with the spontaneous purgation of the physically degenerate couple from Paradise. The narrative proceeds, rather, in a considerably different, far more conventional mode, and this second version of the expulsion begins immediately after the Father's conversation with the Son. The Father now announces his intentions to the entire heavenly community:

> Lest therefore his now bolder hand
> Reach also of the Tree of Life, and eat,
> And live for ever, dream at least to live

---

7. *Works of Winstanley*, p. 115. French, in *New light of Alchymie*, suggests, like Milton, that the expulsion occurs to preserve the "true Elements . . . most pure, temperate, equally proportioned in the highest perfection" (p. 107).

8. Michael Hollington, ed., *Paradise Lost, Books XI–XII* (Cambridge: Cambridge University Press, 1976), p. 56n. Fallon notes suggestively that "there is something of an ontological immune system at work in the poem," in *Milton among the Philosophers*, p. 241.

> For ever, to remove him I decree,
> And send him from the Garden forth to Till
> The Ground whence he was taken, fitter soil.
>
> (11.93–98)

Purged of the hint of mercy and regret that softened the Father's earlier justification of the expulsion, this stern decree posits an entirely new rationale for the deportation of Adam and Eve. Concerned now to assure himself of the first couple's punishment, the Father does not leave justice to a chain of earthly causes; he must rather intervene in the realm of nature to prevent Adam and Eve from indulging in the Tree of Life, the fruit whereof, he freely admits, might allow them to do no more than "dream at least to live / For ever."[9]

In order to forestall further transgression, the Father effects this punitive decree by appointing Michael to expel Adam and Eve:

> *Michael*, this by behest have thou in charge.
>
> .   .   .   .
>
> Haste thee, and from the Paradise of God
> Without remorse drive out the sinful Pair,
> From hallow'd ground th' unholy, and denounce
> To them and to thir Progeny from thence
> Perpetual banishment. Yet lest they faint
> At the sad Sentence rigorously urg'd,
> For I behold them soft'nd and with tears
> Bewailing thir excess, all terror hide.
>
> (11.99–111)

In this scenario of expulsion, the Father acts as judge and pronounces sentence on the sinful pair, actually enjoining an emissary angel to drive the first parents bodily from their home. Far from envisioning the expulsion as a natural, elemental reaction to human sinfulness, this passage expresses the poem's predominant fictive mode of a direct interaction between man and his entirely personal God. Whereas the poem's first, naturalistic explanation of the expulsion asserts a sense of an absolute but impersonal natural justice, the remainder of the epic focuses on the retributive justice of an anthropomorphized deity. In this second, anthropomorphic representation of the expulsion, characterized by the narrative as God's more "official" declaration

9. Milton likewise leaves open in the *Christian Doctrine* the possibility that the Tree of Life may not be simply a "symbol of eternal life," but rather "perhaps the nutriment by which that life is sustained" (15:115).

to his Synod of the Blest, Milton's Father emerges as the stern lawgiver, actively intervening to execute punishment, "judg'd as then best," on a world that seems constantly to require such vigilance.

Between these two versions of the expulsion, between the Father's conversation with the Son and this public sentence of the expulsion among the angels, a vision of natural principles at work in an orderly universe has yielded to an image of the willful and arbitrary execution of justice by a messenger of God. From a certain logical vantage point, the two explanations of the expulsion may not seem so different: the scientific rationale of the inevitability of natural law ("The Law I gave to Nature him forbids") and the voluntary sentence of divine intervention ("to remove him I decree") both point to the Father as the ultimate, first, cause of this event. But although this narrative disjunction may not force us to question ultimate cause, it profoundly affects the way we are to imagine the working out of the secondary principles behind the punishment of the Fall. If we ask of the expulsion the narrator's earlier question, "What cause?" the poem's two explanations provide us with fundamentally conflicting answers. The expulsion is a matter of either logical or juridical consequence: either the first sin results naturally—in accordance with autonomous physical laws—in the spontaneous evacuation of the couple from Paradise, or the fact of that sin requires a transcendent, punitive God to intercede in the realm of nature and forcibly eject the first parents from their home.

As models of Milton's cosmic ontology, these depictions of the expulsion affirm incompatible positions on the implicit laws of agency and organization in God's creation. But despite this rather glaring narrative discontinuity, critics of the last books, nearly all of whom have sought to graft a theological consistency onto the poem, have remained silent about the contradiction. John Broadbent, it is true, has noted in passing the "logical discomfiture" that emerges from the juxtaposition of these divergent accounts.[10] But since the import of the scene is critical to *Paradise Lost*—the expulsion representing nothing less than the loss of Paradise—the considerable disparity between the two narratives deserves, perhaps, more attention than mere dismissal. The "logical discomfiture" that arises from the disparity between God's naturalistic explanation to the Son and his more public statement to Michael is, I believe, rather more than an inconsequential poetic blunder. The confusion reflects in an acute form some of the poem's profoundest contradictions in the philosophy—at once theological, scientific, and political—of agency and organization.

10. Broadbent, *Some Graver Subject: An Essay on "Paradise Lost"* (London: Chatto and Windus, 1960), p. 269.

At issue in the striking disparity between the two versions is the very nature of the agents of change, the operation of the forces of historical cause and effect. And at the center of this narrative contradiction lies precisely that which Milton has intended to justify to men, the ways of God. In the first passage, the Father's evocation of the garden's natural justice bespeaks the vitalist deity we examined in Chapter 4, the liberal God who, having established a set of natural and moral laws that the material world must obey, can now allow his self-governing universe to run quite well on its own. Approaching an animist materialist vision of the autonomous laws of nature, this naturalistic justification recalls the Father's earlier promises of noninterference at the debate in heaven (3.93–130) and at the Creation (7.170–72), the poem's most important statements against the immediacy of God as an efficient cause in the realm of worldly affairs. Although it possesses an absolute moral validity and occurs within the divinely sanctioned realm of nature, the vitalist atmospheric reaction to the sinful couple can in no meaningful way be explained as a divine act of retributive justice.[11]

The conflict that arises between the authoritarian and the liberal narratives of the expulsion of Adam and Eve has one of many organizational analogues in Harvey's conflicting accounts of the movement of the blood. Harvey had accepted the prevailing wisdom that blood was drawn instinctively to the heart, which was, like Paradise, a source of perfection. But how, Harvey asked, are we to explain the expulsion of the blood from the heart? In Harvey's initial, authoritarian formulation of the circulation in 1628, blood must be forcibly expelled from the Edenic heart to effect its circulatory function: if blood is ever to leave the idyllic heart, "where the nourishment is perfected" (p. 94), "and move against its will, and enter into places narrower and colder," then "it has need of violence and an impulsor, such as the heart is only" (p. 95). In 1628, sanguineous expulsion to the cold and narrow reaches of the outer body can occur only through impulsion by a higher power. But in Harvey's reorganized thesis of 1649, it is "the innate heat in the blood it self" that effects its expulsion from the heart, a swelling up of blood "not caus'd by any external agent, but by the regulating of Nature, an internal principle" (p. 187). Milton's depictions of the evacuation of Eden, I propose, reproduce the same organizational outlines of the paradigmatic narratives of causation established by Harvey.

But the period's revolutionary culture produced numerous examples of the conflicting models of causation explored in *Paradise Lost*. The ontological distinction between the two models of agency and organization dramatized

---

11. See Fowler's note to this passage in *John Milton: Paradise Lost*, p. 984.

in the expulsion narratives emerges in a more recognizable theological form in Milton's own *Christian Doctrine*. As a theologian, Milton articulates this distinction in modes of action as a difference between two types of providence, the "ordinary" and the "extraordinary." His definition of "extraordinary providence" imputes volition to the Deity, reflecting the more familiar orthodox understanding of divine providence that we find in the second, public statement on the expulsion: "The extraordinary providence of God is that whereby God produces some effect out of the usual order of nature, or gives the power of producing the same effect to whomsoever he may appoint. This is what we call a miracle. Hence God alone is the primary author of miracles, as he only is able to invert that order of things which he has himself appointed" (15:95). Milton is loath in the theological tract to relinquish the verisimilitude of the miracles recorded in Scripture. "God," he maintains, "decreed nothing absolutely" (14:65) and can therefore "invert the order of nature" at whim. Informing the enactments of "extraordinary providence" throughout the epic, this orthodox assertion of the possibility of miraculous intervention provides the doctrinal foundation for the mythopoeic account of God's instructions to Michael to descend to earth and hasten the pair from the garden.

Under the heading of "ordinary providence" we find Milton's definition of that irrevocable law that, as we have seen, the God of *Paradise Lost* has given to nature: "His ordinary providence is that whereby he upholds and preserves the immutable order of causes appointed by him in the beginning." We can see here how Milton's notion of ordinary providence, his accommodation of the vitalist conception of natural process to a standard principle of Christian doctrine, could begin to take on the appearance of a natural realm almost secular in its autonomy. Milton the theologian, however, assumes the voice of Abdiel and hastens to dispel any of the atheist overtones produced by this ordinary, naturalized providence: "This is commonly, and indeed too frequently, described by the name of nature; for nature cannot possibly mean anything but the mysterious power and efficacy of that divine voice which went forth in the beginning, and to which, as to a perpetual command, all things have since paid obedience" (15:93). Milton insists, as we have seen him insist in his accounts of Creation, that the providential force known as "nature" received its original empowerment by God. But since its inception by the divine voice, the law of nature is nonetheless said to hold perpetual sway over all things. Despite Milton's stated refusal to equate ordinary providence with "nature," the logic of his argument still corresponds to the gesture of the many seventeenth-century materialist philosophers whose work had the radical effect of reducing "divine providence to the original scheme of crea-tion."[12]

In *Paradise Lost*, Milton is perhaps less hesitant than he is in *Christian Doctrine* to suggest an equation between God and natural law. When Abdiel explains to Satan that "God and Nature bid the same," he can be construed to identify God, in effect, with the rules of the self-regulating natural uni-verse that constitute ordinary providence. He defines the God of *Paradise Lost* not as the anthropomorphic figure of the Old Testament but as a figure closer to the absolute, rationalistic principle envisioned by marginal theolo-gians as diverse as the Cambridge Platonists, on the right, and, on the left, the Diggers.[13] Central to the poem's material narrative is the physiological depiction of the effects of the Fall. In his *Christian Doctrine*, Milton asserts the orthodox view that God's decision to forbid the fruit of the Tree of Knowledge was wholly arbitrary; because the fruit was "in its own nature indifferent" [*neque bonum . . . neque malum*] (15:112–13), God forbade its consumption for no other reason than to test the obedience of Adam and Eve. In *Paradise Lost*, however, Milton subjects God's dicta to a higher standard of logic and justifiability. Laboring to render both substantial and reasonable God's proscription of the fruit, Milton follows the vitalist reading of the prohibition established by van Helmont, for whom "Death was placed in the Apple, but not in the opposition of eating: And therefore that Death from the eating of the Apple was natural, being admonished of, but not a

12. Richard Westfall, *Science and Religion in Seventeenth-Century England* (New Haven: Yale University Press, 1958), p. 86. Many seventeenth-century writers on providence, seeking to distance themselves from Calvinist voluntarism, revived the medieval distinction between "ordinary" providence and its "extraordinary" counterpart. The difference between these modes of divine action was just as often expressed with the terms "general" and "special," "habitual" and "actual," or "mediate" and "immediate." This semantic gesture of differentiation, however, accrued a tremendous variety of meanings, and only on occasion did theologians parallel Milton in relating the "ordinary" form of providence to natural law. For a brief survey of the range of meanings these distinctions expressed, see Worden, "Providence and Politics," pp. 60–61.

13. Ralph Cudworth, for example, concludes his formidable *True Intellectual System of the Universe*, written in the 1650s and 1660s, with an argument for the subordination of God to an absolute principle of justice: "The *Right* and *Authority* of *God* himself, who is the *Supreme Sovereign* of the *Universe*, is also in like manner *Bounded* and *Circumscribed* by *Justice*. God's *Will* is *Ruled* by his *Justice*, and not his *Justice Ruled* by his *Will*; and therefore God himself cannot *Command*, what is in its own nature *Unjust*" (*The True Intellectual System of the Universe* [London, 1678], p. 897). For a discussion of the more radical, leftist identification of God with reason, see G. E. Aylmer, "The Religion of Gerrard Winstanley," in *Radical Religion in the English Revolution*, ed. J. F. McGregor and B. G. Reay (Oxford: Oxford University Press), pp. 96–98. For the argument that Milton represents his God as just such a principle of rational justice, see Saurat, *Milton*, pp. 113–33; Barker, *Milton and the Puritan Dilemma*, pp. 326–28; and Andrew Milner, *John Milton and the English Revolution* (London: Macmillan, 1981), pp. 115–18.

Curse threatned by a Law."[14] With a description of the first couple's physiological reaction to the fruit, Milton establishes the possibility that mysterious organic powers were lurking in the fruit all along.[15] He depicts the gradual physical deterioration of Adam and Eve as the natural consequence of their ingestion of a pathogenic meal:

> Soon as the force of that fallacious Fruit,
> That with exhilarating vapor bland
> About thir spirits had play'd, and inmost powers
> Made err, was now exhal'd, and grosser sleep
> Bred of unkindly fumes, with conscious dreams
> Encumber'd, now had left them, up they rose
> As from unrest, and each the other viewing,
> Soon found thir Eyes how op'n'd, and thir minds
> How dark'n'd.
>
> (9.1046–54)

The description of the drunkenness that follows the ingestion of the fruit, the play of vapors about their intellectual spirits which seems irrevocably to darken their minds, attest to Milton's attempt to justify the effects of the Fall by an appeal to material principles.[16] By making the conventionally "arbitrary" interdiction against the fruit explicable on natural grounds, Milton's physiological images have the radical effect of identifying and collapsing God's law with the law of nature; whereas the fruit as arbitrary restraint necessitates God's direct punishment of man's disobedience, a pathogenic fruit occasions the Fall without any additional intervention from above. Milton's naturalization of divine law thus holds two primary consequences for the poem's justification of historical cause. The homology between God's purposive decrees and the ordinary providence of pre-established natural law

14. Van Helmont, *Oriatrike*, p. 653. Turner discusses the Helmontian naturalization of the prohibition in *One Flesh*, p. 150.

15. See Empson, *Milton's God*, p. 188; and Kerrigan, *Sacred Complex*, p. 252. Michael Lieb, in *Poetics of the Holy: A Reading of "Paradise Lost"* (Chapel Hill: University of North Carolina Press, 1981), pp. 105–8, argues against any naturalist reading of the fruit's significance.

16. Kerrigan adds that not only Adam and Eve but the Earth, too, suffers from the "digestive" processes of nature: "Representations of digestive illness surround the crisis. At each disobedience nature suffers the cramps of a bellyache, 'Sighing through all her Works' (9.783) and trembling 'from her entrails' (9.1000). These macrocosmic symptoms in the body of mother earth hailed the arrival of a pathogenic nature, corrupt with flatulence: 'Vapor, and Mist, and Exhalation hot, / Corrupt and Pestilent' (10.694–695)" (*Sacred Complex*, p. 252). Lieb reviews the natural alchemical constitution of Milton's cosmos in his appendix "Creation and Alchemy" in *Dialectics of Creation*, pp. 229–44.

absolves the Deity from the cruel, punitive streak revealed in his expulsion of the first couple. But this identification of God with nature also works to diminish the Father's status as an independent being in whose hands rest the guidance and the judgment of human affairs. Milton's Father asserts that at the final judgment he will bequeath to his Son all divine authority. But long before that ultimate abdication when God shall be "All in All," the Father seems already to have ceded control to a blind and impersonal order of nature with which man is allowed to interact alone.

As we have seen from the competing narratives of the expulsion, the epic has inscribed in its theodicy two strands of agential philosophy that can be reconciled only with considerable hermeneutic finesse. The striking oppositions that arise in the poem between the impersonal, naturalistic form of explanation and that of God's retributive justice reflect a significant tension between two conceptions of the divine role in human history, attitudes toward history that are grounded in radically different organizational ontologies. In his *Christian Doctrine*, as we have seen, the Ramist framework allows Milton to juxtapose the forces of ordinary and extraordinary providence as two logically related modes of divine control: he holds tight to the idea of God's absolute freedom and omnipotence even as he forwards the premise of the irrevocable laws of nature. It is precisely the insoluble conflict between the concepts of divine omnipotence and irrevocable natural law that provides the source for some of the most fundamental inconsistencies in Milton's poem. The God of *Paradise Lost* is at once an impersonal embodiment of reason and an emotive, willful executor of justice: at one point, for example, Milton's Father expresses his impersonal identification with a larger moral structure as he states his inability to "revoke the high Decree / Unchangeable, Eternal" (3.126–27), but he later flexes his arbitrary will like the God of Calvin and boasts that his "goodness" is "free / To act or not, Necessity and Chance / Approach not mee, and what I will is Fate" (7.171–73).[17] In spite of a sanguine denial of theological inconsistency in *Christian Doctrine*,[18] the narrative and theoretical disjunction that occurs between the two expulsion passages in *Paradise Lost* exposes a worried fear for the irreconcilability

17. On the inconsistency of Milton's views on divine freedom, see Mili N. Clark, "The Mechanics of Creation: Non-contradiction and Natural Necessity in *Paradise Lost*," *English Literary Renaissance* 7 (1977): 223. Kendrick, in *Milton*, p. 115, describes as "embarrassing" the "eerie co-residence within [*Paradise Lost*] of doctrinal and psychological motivations and justifications." See also Stephen Fallon, " 'To Act or Not': Milton's Conception of Divine Freedom," *Journal of the History of Ideas* 49 (1988): 425–52.

18. "A certain immutable necessity of acting rightly, independent of all extraneous influence whatever, may exist in God conjointly with the most perfect liberty, both which principles in the same divine nature tend to the same point" (Milton, *Works*, 14:73).

of these two forms of providential agency. This disjunction exposes further, I believe, the text's inscription in a larger cultural conflict of the revolutionary years, the struggle for an authoritative discourse of agency and organization.

This discursive crisis hinged largely on the contested notion of the "law of nature." Throughout the century one could assert with little resistance the superiority of a divinely sanctioned "law of nature" over a law or set of principles merely human in origin. By midcentury, however, this distinction was subject to a number of refinements, and it became increasingly acceptable to assert the affinity, and sometimes the equality, of a law of nature *tout court* with a law of God. The principle of the law of nature became susceptible to infinite invocations in the struggle to overturn social and political authorities that had only recently enjoyed the unchallenged prestige of divine sanction. The usurpation of a traditional, perhaps even divinely sanctioned, law by a law authorized now by its origins in "Nature" became one of the period's most vivid emblems of revolution. But the political applicability and the ideological implications of the revolutionary turn to natural law were never neatly demarcated, and versions of the antinomy between natural and divine law formed the central paradigm for a wide variety of intellectual disciplines, practices within discursive communities that, if they did not cohere into an identifiable socioeconomic class or political party, nonetheless shared a politically consequential epistemic strategy.

This paradigmatic appeal to natural law proved most useful in the struggle against power whose authority rested in a tradition of arbitrary rule. In this age of revolution, it was called on preeminently, and most dramatically, by Parliamentarians concerned to subject the prerogative rule of the English monarch to a "law of nature" articulated as a guarantor of individual rights. But it also supplied the conceptual leverage for other groups of varying degrees of relation to the broad concerns of Parliament: the "law of nature" was invoked by Levellers resisting the state regulation of the economy in their push for free trade, and by many tolerationist Independents who struggled during the 1650s against the Presbyterian stranglehold on the church.[19] These palpable and explicitly political movements were accompanied by impassioned acts of resistance in discursive spheres no less important to English men and women. The tool of the "law of nature" proved instrumental for theologians, both Anglican and nonconformist, attempting to force the arbitrary and omnipotent God of Calvin to obey the moral "laws of nature" established at the Creation; and for natural philosophers determined to force

---

19. J. P. Cooper, "Social and Economic Policies under the Commonwealth," in *The Interregnum: The Quest for Settlement, 1646–1660*, ed. G. E. Aylmer (London: Macmillan, 1982), p. 124.

the miraculous phenomena recorded in Scripture into conformity with a "law of nature" imaged as the unswerving regularity of physical process. In whatever sphere its authority was imagined to extend—in political philosophy, economics, theology, or natural philosophy—the propounding of a "law of nature" functioned invariably as a *constraint* on the traditional discourses of authority. English intellectuals asserted for the first time, and with surprising volubility, the freedom of certain realms of action from arbitrary intervention. Carving out a broad conceptual system we have identified as a proto-liberalism, the varied and often conflicting proponents of natural law labored to confer autonomy on what they imagined to be self-sufficient organizations of agents and objects—the self-enclosed economies of everything from political representatives and religious sects to marketable commodities and physical forces.

To the extent Milton identifies the God of *Paradise Lost* with the necessary laws of universal order, he participates in a broad discursive movement consonant with the nascent liberal drive toward philosophies of decentralized organization that championed the value and primacy of natural law. In the theological sphere, Milton is following the specific example of the early-seventeenth-century Dutch theologian Jacob Arminius. As the great proponent of free will, Arminius grounded his liberation theology on the confinement of God to an abstract principle of justice: "God is good by a natural and internal necessity, not *freely*."[20] We have already seen how in Milton's time this Arminian notion of the necessity of divine goodness could be displaced onto a more scientific understanding of divine necessity, the less reverent identification of God with an irrevocable natural law. But whether God was identified with a principle of goodness or a principle of physics, the doctrine of divine necessity served a crucial function: it provided an important psychological sanction for the belief in its logical converse, the free will of man. Because the principle of an unimpeachable realm of ordinary providence works to limit the capacity of the Deity to rupture natural law and intervene in the affairs of men, Milton's naturalism guarantees the effectiveness of rational human agency; it leaves man free to exercise his reason on the identifiable patterns of natural and moral consequence.

Milton's establishment of a vitalist discourse of natural law is not, however, confined to what may seem the rather sterile theological question of free will. The alternative, naturalist narrative we have charted is at the heart of the poem's attempt to ground in the authority of science a new logic of individualism, a principle of free agency paradoxically, but crucially, contin-

---

20. *The Writings of James Arminius*, trans. James Nichols and William R. Bagnall, 2 vols. (Grand Rapids, Mich.: Baker Book House, 1956), 1:344–45.

gent on the stable existence of unswerving, unbreakable laws. The first proponents of free-trade economics were arguing as early as the 1620s that a
merchant can prosper only when he can depend on the predictable operation
of the autonomous laws of the marketplace; there became an increasingly
strong conviction that a merchant can calculate value to his advantage only
in an economy free from interventionist practices such as monarchic price-
fixing and the granting of monopolies.[21] As Jürgen Habermas has argued,
the emergence of a "public sphere" made up of autonomous, independent
agents depends, as in the emergence of the liberal marketplace, on the existence of impersonal "universal rules."[22] It is a logic identical to this faith
in the benefits of an autonomous market for the exchange of goods which
structures Milton's consideration of the liberal economy of moral life. Only
in a world in which the basis of goodness and justice is exempt from the
arbitrary determinations of a capricious God can it be imagined that individuals can freely employ their innate powers of reason as effective agents in
their own spiritual, and political, prosperity.

We must not, of course, lose sight of the fact that Milton's story of the
interdependence of human freedom and divine constraint, though a substantial aspect of the poem, halts abruptly at Michael's lengthy recitals of
God's interventions in the course of history. Just as Milton's scene of the
induction of Michael to enforce the Father's *congé* is itself an assertion of
the strong arm of providential control, so Michael himself envisions history
as a forum for God's retributive justice. Asserting a pattern of human sin
and divine punishment that runs counter to Milton's naturalistic perspective,
Michael emerges as the poem's premier spokesman for the theocentric theory of history and change. The heavenly Father whom Michael depicts is a
"God who oft descends to visit men / Unseen . . . to mark thir doings"
(12.48–50), a God who laughs with malice "to see the hubbub strange / And
hear the din" of the builders of Babel, on whom he has placed a wicked
curse (12.60–61). As he rehearses in schoolbook abstractions the patterns of
"supernal Grace contending / With sinfulness of men," Michael affirms a
disappointingly orthodox faith in the miracles of Scripture. As he teaches
God's direct exercise of justice on the lives of men, he embarks on a re-

21. See Appleby, *Economic Thought and Ideology*.
22. Jürgen Habermas, *The Structural Transformation of the Public Sphere: An Inquiry into
a Category of Bourgeois Society*, trans. Thomas Burger (Cambridge: MIT Press, 1991), p. 54:
"These rules, because they remained strictly external to the individual as such, secured space
for the development of these individuals' interiority by literary means. These rules, because
universally valid, secured space for the individuated person; because they were objective, they
secured a space for what was most subjective; because they were abstract, for what was most
concrete."

lentless delineation of an authoritarian philosophy of organization that nearly overwhelms the poem's attempt to engender a discourse of liberal individualism.

The Tyranny of Divine Retribution

One of the poem's most powerful accounts of God's direct intervention in human history can be found in Michael's famous justification of political tyranny, an account that perhaps best exposes the lines of intersection between Milton's poetic historiography and the political history of revolutionary and Restoration England. In the description of the terrible emergence of Nimrod, Milton lays bare in the starkest terms imaginable the political philosophical consequences of the theology of extraordinary providence.[23] Michael's depiction of man's subjection to tyrants, it is true, begins with the liberal, secular analysis Milton had forged in the regicide tracts of the Vitalist Moment. Calling on the conventional analogy between man's moral nature and the structure of the political state, Michael initially attributes tyrannical rule not to divine will but to sinful human desire:

> yet know withal,
> Since thy original lapse, true Liberty
> Is lost, which always with right Reason dwells
> Twinn'd, and from her hath no dividual being:
> Reason in man obscur'd, or not obey'd,
> Immediately inordinate desires
> And upstart Passions catch the Government
> From Reason, and to servitude reduce
> Man till then free.
>
> (12.82–90)

Although Michael grants fallen man, whose faculty of reason might actually be obscured, less credit for the capacity for reason than Milton might have in an earlier work such as *Areopagitica*, he amplifies one of Milton's standard narratives of undisciplined behavior. The Fall either diminishes man's faculty of reason or encourages him to disobey it; thus dissipated, fallen man allows passion to rule over reason, an unlawful usurpation that inevitably reduces his freedom by diminishing his rational capacity for virtuous choice. So far, Michael's analysis of lost liberty does not radically depart from Milton's

---

23. See Patterson's discussion of the importance of the figure of Nimrod for the period's political philosophical understanding of monarchy, in *Reading between the Lines*, pp. 252–55.

highly liberal conception of justice: it focuses on the self-inflicted loss of freedom that results from the damage to reason after the Fall, and it locates the origin of a disastrous political organization in the liberal realm of individual choice.

As Michael continues his explanation, however, he quickly lifts his attribution of cause from this human, psychological plane to the supernal stratum of the anger of a vengeful God:

> Therefore since hee permits
> Within himself unworthy Powers to reign
> Over free Reason, God in Judgment just
> Subjects him from without to violent Lords;
> Who oft as undeservedly enthral
> His outward freedom: Tyranny must be,
> Though to the Tyrant thereby no excuse.
> Yet sometimes Nations will decline so low
> From virtue, which is reason, that no wrong,
> But Justice, and some fatal curse annext
> Deprives them of thir outward liberty,
> Thir inward lost.
>
> (12.90–101)

Nations can grow so sinful, explains Michael, that it is not enough that they punish themselves with a loss of inward freedom: only the execution of divine "Justice" or "some fatal curse annext" by an external power can properly mete to a people the punishment it deserves. Michael later relates how the Israelites, "in thir earthly *Canaan* plac't / Long time shall dwell and prosper"; but, he adds, compelled to amplify his theological voluntarism, "when sins / National interrupt thir public peace, / Provoking God to raise them enemies," the Father must intervene to ensure the proper exercise of judicial punishment (12.315–20). In consummate "Judgment just," therefore, Michael's God willfully subjects man to the scourge of earthly tyrants, a providential act that validates for Michael his chilling apothegm of Calvinist necessity: "Tyranny must be."[24]

Michael has introduced a theology so counter to the prevailing image of

24. Michael reproduces in this speech the orthodox Calvinist justification of undesirable political organizations: "The authority possessed by kings and governors over all things upon earth is not a consequence of the perverseness of men, but of the providence and holy ordinance of God, who has been pleased to regulate human affairs in this manner" (Calvin, *Institutes of the Christian Religion*, trans. John Allen, 2 vols. [Philadelphia: Westminster Press, 1935], 2:636). On Calvin's justification of political tyranny, see Walzer, *Revolution of the Saints*, pp. 35–38.

Milton's political liberalism that he has forced Milton's readers into blind misreadings of his speech. Even the subtlest critics of the last books, laboring to humanize Michael's lessons, have glossed over these unsettling instances of intervention in what is surely one of the poem's central moments of historical explanation. Thus, in attempting to naturalize Michael's account of a highly unnatural chain of cause and effect, Barbara Kiefer Lewalski writes simply that "political tyranny *has its roots* in the loss of true government over the self," and Mary Ann Radzinowicz explains that the "deprivation of outer freedom *is the inevitable result* of the surrender of inner freedom" (italics mine).[25] Michael, however, has in no way articulated so logical a chain of cause and effect; he quite clearly attaches the link of arbitrary divine will to this chain of inevitability, concluding his history of political subjection by rehearsing the pattern of human sin and divine punishment that had become in Milton's time a commonplace of Calvinist theological authoritarianism. Augustine, convinced of man's depravity and his absolute incapacity for free moral choice, had described political enslavement as a "humiliation visited on the conquered by divine judgement, either to correct or to punish their sins"; and Calvin had actually nominated the earthly monarch as "an executioner of God's wrath."[26] Since Scripture had already provided Milton with a similarly nominalistic account of divine punishment, as well as with the figure of the vengeful God, Michael's doggedly voluntaristic emphasis on divine retribution may seem unavoidable. But the truly puzzling aspect of Michael's providential justification here, one that critics consistently overlook, lies in its hostility to so many of Milton's most fundamental organizational tenets. Michael, in describing tyranny as the result of God's voluntary punishment of man rather than a concession to man's mistaken but voluntary choice, forwards an explanation that contradicts some of Milton's most impassioned liberal etiologies of political organization.

In order to understand the implications of the shift in Michael's political analysis to an arbitrary, punitive God, we must return to the first formulations of Milton's philosophy of political organization, the regicide treatises of 1649 and 1651. It is here, in *Tenure of Kings and Magistrates* and the first *Defence of the English People*, that Milton carves out a liberal politics from the larger discourse of decentralized organization emerging, as we have seen, across a range of disciplines during the Vitalist Moment. In Milton's etiology of the proper government of popular sovereignty—the rule of a constitu-

---

25. Lewalski, *"Paradise Lost,"* p. 262; Radzinowicz, "Politics of *Paradise Lost,"* p. 228.
26. St. Augustine, *The City of God,* trans. Henry Bettenson (Harmondsworth: Penguin, 1984), p. 875; John [Jean] Calvin, *Commentaries on the Epistle of Paul the Apostle to the Romans,* trans. J. Owen (Edinburgh: Calvin Translation Society, 1849), p. 481.

tional monarch elected by the people—a tyrannized populace had asserted its power to subject the arbitrary monarch to a higher set of rational laws:

> Then did they who now by tryal had found the danger and inconveniences of committing arbitrary power to any, invent Laws either framed, or consented to by all, that should confine and limit the autority of whom they chose to govern them: that so man, of whose failing they had proof, might no more rule over them, but law and reason abstracted as much as might be from personal errors and frailties. While as the Magistrate was set above the people, so the Law was set above the Magistrate. (5:9)

This articulation in *The Tenure of Kings and Magistrates* of the ideal polity, governed now by a set of laws "abstracted as much as might be from personal errors and frailties," clearly supplies Milton with the organizational framework for his later, theological doctrine of ordinary providence: God may be set above his creatures, but the laws God gave to nature have been set above God, constraining him to govern his creation as an abstract, impersonal principle of law and reason. Although Milton does not formulate the theological version of this subjection to law until his composition of the *Christian Doctrine*, the interrelated logics of a circumscribed ruler and a circumscribed God come to structure all his considerations, in these early political treatises, of the origins, the causes, of political organizations. His commitment to a liberal economy of cause and effect forces him, in these works, to ascribe tyranny not to divine will but to the secular principle of individual human choice.[27] Milton devotes an entire chapter of the *Defence of the English People* to his contradiction of Salmasius's seemingly overwhelming evidence of God's agency in the establishment of tyrants. "God has never been so hostile to the whole human race," he concluded there, that he would impose tyranny: "God has decided then that the form of a commonwealth is more perfect than that of a monarchy as human conditions go, and of greater benefit to his own people. . . . He granted a monarchy only later at their request and then not willingly."[28] For the Milton of the Vitalist Moment, who was driven to justify all political organizations on secular grounds, the

27. See Martin Dzelzainis, who explains in his edition of Milton's *Political Writings* (Cambridge: Cambridge University Press, 1991) that Milton "is at pains to distance himself and the Army from the allegations of religious enthusiasm and zealotry levelled by the Presbyterians. His scepticism about divine commands, his dissent from voluntarism, and his emphasis on reason all stem from the need to fashion a less vulnerable, because more secular, kind of argument" (p. xv).

28. Milton, *Political Writings*, ed. Dzelzainis, p. 80; I cite this edition as a clearer translation than S. L. Wolff's in *Works of John Milton*, 7:75, 77. See the related argument in the earlier regicide tract, *Tenure of Kings and Magistrates*, 5:1.

logic of the commonwealth depended on an analogous logic of a rationalized God, a deity who may, if pressed, grant his people a monarchy, but only "to show that He had left the people their choice to be governed by a single person" (7:77). The Milton of this period would have been constitutionally incapable of representing a God who set up a tyrant as an unsought punishment. In the causal chain enunciated in these most liberal, humanistic political genealogies, there is simply no logical link that can accommodate Michael's identification of the tyrant as a scourge of God.

There can be little question that Michael's depictions of an intervening God damage the poem's theoretical assertions of a retiring, laissez-faire divinity. We can say, further, that his reliance on an implicitly voluntarist theology of divine agency also begins to disrupt the logic of the republicanism that forms the thematic content of his poem's own political analysis. At the very moment Michael forwards this stern political lesson on the ideal of "fair equality, fraternal state" (12.26), his reliance on a peculiarly authoritarian theology weakens the force of his commitment, in the political realm, to egalitarianism. Given the scope of Milton's investment in the organizational imperative, I think we can see a disturbing way in which the authoritarian ruler, Nimrod, finds a rhetorical affirmation of his arbitrary rule in Michael's representation of an arbitrary God, who, like Nimrod himself, has resisted all subjection to the laws of nature, "to himself assuming / Authority usurped" (12.65–66). Milton's God, of course, punishes Nimrod for his despotic presumption, setting on the tongues of his builders "a jangling noise of words unknown" (12.55). But Milton adds to this image of justifiable punishment an act of divine pleasure mentioned nowhere in his Genesis original: "great laughter was in Heav'n" (12.59). Michael insists, on the level of argument, on God's antagonism to such political presumption, but, on the level of figure, Milton's clamorous insertion of an image of God's whimsical menace exposes the mutually constitutive relation of a voluntarist theology and a politics of absolutism.[29]

The most visible sign of this fissure in the organizational integrity of the political philosophy of the last books is the representational conflict we have observed between the two versions of the expulsion of Adam and Eve. The anthropomorphic God so carefully delineated cannot be easily reconciled to the God of vitalist science. Conversely, the image of the naturalistic deity poses its own threat to the poem's integrity as a story. When William Empson engaged approvingly Denis Saurat's argument that Milton's God was little

---

29. Milton, as Fowler notes, in *John Milton: Paradise Lost*, p. 612, imports into this story the figure of God's laughter from Psalms 2:4. Radzinowicz defends the representation of God's grim humor in *Toward "Samson Agonistes*," p. 304n.

more than a rational, pantheistic principle, he conceded that this "interpretation of the epic makes nonsense of most of its narrative."[30] And, indeed, the notion of an impersonal deity seems to question the intended truth-value of much of the poem's anthropomorphic mythopoesis. We must now ask why the poem contains such radically conflicting discourses of divine agency and cosmic organization. What is the relation between the anthropomorphic history Michael expounds and the more naturalistic sense of causation that surfaces elsewhere in the poem? What form of justification may we as critics employ in accounting for the explanatory strategies so divergent that they threaten the poem's narrative integrity? I propose that we examine the possibility that this conflict between rationalistic and anthropomorphic conceptions of the Deity, a tension so familiar that it was essentially constitutive of the theological debates among Milton's contemporaries, is submitted to a poetic—though not, ultimately, a logical—resolution within the poem itself. Committed to the project of theodicy, Milton, I think, strives to master this central problem in organizational ontology, and the last books of the poem show signs of an attempt, however inadequate, to resolve some of the contradictions threatening his narrative representation of agency and organization.

The Better Covenant of Natural Law

The angel Raphael, as we saw in Chapter 4, complicated his role as "Divine Historian" by engaging discursive systems founded on conflicting philosophies of organization. In the final books of *Paradise Lost*, Milton's epic narrator may portray, as in the first account of expulsion, the vitalist God we recognize from Raphael's story of Creation, but the angel Michael, who bears by far the greatest narrative burden in these books, is committed to a consistent, perhaps even unyielding, representation of God as the arbitrary God of Calvin. I want here to investigate the poem's figuration of Michael's function in the larger narrative. The identification of this function as pedagogical, rather than faithfully historiographical, should enable us to contextualize the alarming authoritarianism that comprises the conclusion of Milton's great poem of Christian liberty.

Although Michael is ostensibly instructed to cushion the blow of God's wrath, Milton does not shrink from depicting him throughout the poem as an imposing, even terrifying, figure. On both the discursive and the narrative levels, the fierce warrior angel embodies in the epic precisely that principle

---

30. Empson, *Milton's God*, p. 144. See Saurat, *Milton*.

that his history celebrates—the swift, willful execution of divine justice. When we first encounter Michael in Book Two, he has already acted on God's behalf to instill a painfully learned moral lesson: at the time of the poem's first expulsion—the ejection of the rebel angels from heaven—Michael's violent actions induced among Satan's league a powerful fear of further war, "so much the fear / Of Thunder and the Sword of *Michael* / Wrought still within them" (2.293–95). Milton reminds us of this awful and guilt-inspiring power of Michael's sword when the Father gives Michael the immediate, practical instructions to preserve the Tree of Life by placing "Cherubic watch, and of a Sword the flame / Wide waving, all approach far off to fright" (11.120–21). And, finally, it is Michael's sword of conscience we find at the poem's close, in the touching moment when the remorseful Adam and Eve, "looking back," behold the startling vision of the "flaming Brand, the Gate / With dreadful Faces throng'd and fiery Arms" (12.643–44). Michael, with his terrifying sword, induces a fear of punishment that savagely enforces adherence to the law.

This effect of Michael's sword in inciting obedience through fear should provide us with a useful key to his work as prophetic historian. As it is God's willful punishment of man's wickedness that, at least from Michael's perspective, initially prompts historical change, so it is nothing but a fear of divine punishment that characterizes man's first attempts at self-government. Members of the postdiluvian tribe of Noah, for example, will respond to God's punishing flood and mend their behavior "while the dread of judgment past remains / Fresh in thir minds, fearing the Deity" (12.14–15). This primitive form of self-government we have already encountered in *Paradise Lost*. After his narrative of the Son's punishment of the rebel angels, Raphael advises Adam to "remember, and fear to transgress" (6.912). And it is the memory of God's threatened punishment that Adam recalled so vividly before the Fall:

> Sternly he pronounc'd
> The rigid interdiction, which resounds
> Yet dreadful in mine ear, though in my choice
> Not to incur.
> (8.333–36)

The events of Book Nine of *Paradise Lost*, of course, prove in short order that the mere fear of arbitrary punishment, for all its dreadful resonance, is not in itself a sufficient ethical guide. Adam's fear of an arbitrary power is scarcely more elevated than the fear Hobbes hoped would be aroused by the ideal sovereign of his authoritarian utopia, that *"Mortall God"* who "hath use of so much Power and Strength conferred on him, that by terror thereof, he is enable to forme the wills of them all."[31] As the poem makes clear in the event of the Fall, the most striking example of the ineffectiveness of sheer terror, Michael's deployment of an authoritarian theology to frighten and intimidate Adam can in no way be construed as an adequate, liberal education.

We have established Michael's use of a nominalist providentialism, like the sword with which he inculcates this unswervingly centralized organizational philosophy, as a sharp instrument of moral pedagogy. But we have yet to understand the relationship between this anthropomorphic vision of divine history that Michael displays and the more vitalist sense of natural historical agency that surfaces elsewhere in the poem, the dialectic, in short, of the law of God and the law of nature. In order to clarify Milton's attempt to bring into focus the status of these two organizational modes, I propose that we return to the scene of Michael's induction. The Son of God, with a flourish that befits such an important occasion at the heavenly court, summons the angels to the ceremony of Michael's initiation:

> he blew
> His Trumpet, heard in *Oreb* since perhaps
> When God descended, and perhaps once more
> To sound at general Doom.
>
> (11.73–76)

Milton's elaboration on the possible subsequent history of the Son's trumpet associates Michael's induction with Moses' administration of law and judgment: it is Mount Oreb, or Sinai, on which God delivers the Ten Commandments to Moses. We hear the Son's trumpet again in Michael's account of God's thunderous dispensation of Mosaic law:

> God from the Mount of *Sinai*, whose gray top
> Shall tremble, he descending, will himself
> In Thunder, Lightning and loud Trumpet's sound
> Ordain them Laws.
>
> (12.227–30)

Milton carefully amplifies this image of the trumpet of law and justice to announce the nature of the lesson of divine judgment Michael will teach— both the harsh lesson of the actual expulsion and the history of God's au-

---

31. Hobbes, *Leviathan*, p. 227.

thoritarian dispensation of divine rewards and punishments. But the image of the trumpet also signals the narrative status of the pedagogical function Michael will serve for the rest of the poem: the rigorously typological figuration forces here a recognition of Michael as a prototype of Moses, the deliverer of divine law to the Israelites. Acting the stern prophet and lawgiver as he delivers the laws of God that discover sin, Michael speaks to Adam as Moses will later to the Hebrews.[32]

In order to clarify what I see to be Milton's own attention to the problem of Michael's theology, let us turn to the passage in the narrative of Moses that articulates the status of the pedagogical function that Michael's Mosaic authoritarianism is intended to serve. In this famous enunciation of Milton's typological schema, Michael teaches Adam about the historicity of Mosaic teaching, its ultimately limited position on the completed course of human history:

> So Law appears imperfet, and but giv'n
> With purpose to resign them in full time
> Up to a better Cov'nant, disciplin'd
> From shadowy Types to Truth, from Flesh to Spirit,
> From imposition of strict Laws, to free
> Acceptance of large Grace, from servile fear
> To filial, works of Law to works of Faith.
>
> (12.300–306)

Most critics have interpreted the movement "From shadowy Types to Truth" to refer only to the long race of time that is Christian history.[33] But in employing the term "disciplin'd" to mark the historical passage from types to truth, Milton signals an analogy between that larger historical movement and what is surely a more familiar sense of the process of "discipline," the moral education Michael is bestowing on Adam. In its evocation of the traditional humanist belief that historiography—what we call today the discipline of history—is an important rhetorical tool for moral edification, the word "discipline" establishes the parallel between the rectilinear history Michael relates and the angel's own place in the linear process of Adam's education, a process that will ideally release Adam from the Hobbesian state of "servile fear" that Michael's authoritarian history seems to have induced.

32. Critics who have suggested a quasi-typological connection between Michael and Moses include Leslie Brisman, *Milton's Poetry of Choice and Its Romantic Heirs* (Ithaca: Cornell University Press, 1973), p. 188; and John T. Shawcross, *With Mortal Voice: The Creation of "Paradise Lost"* (Lexington: University Press of Kentucky, 1982), p. 128.

33. The theory of typology sketched in this passage is explained most notably by William G. Madsen, *From Shadowy Types to Truth* (New Haven: Yale University Press, 1968).

Through this juxtaposition of Michael's narrative of future history with the education of Adam, Milton situates the orthodox vision of the race of time as a stage in the familiar process of humanistic education.

Like the organizational shift between Harvey's two accounts of the blood's circulation, this disciplinary process is one of systemic liberalization. It is intended to move man, considered both phylogenetically and ontogenetically, "From imposition of strict Laws, to free / Acceptance of large Grace" (12.304–5), a movement away from divine intervention and toward a world of human choice and voluntary acceptance of help from above. Although these lines imply a theological consistency through the syntactical parallel between "imposition" and "acceptance," the implicit grammatical agent behind each activity has shifted from God to man, from the *divine* imposition of laws to the *human* acceptance of divine grace. This new anthropocentric perspective has the radical effect of emptying the very notion of grace of its Calvinist selectivity and dependence on an arbitrary God. Although Milton here, as throughout *Paradise Lost*, saves the appearance of grace, the poetics of his theology suggests a certain logical slip into the Pelagian belief in a human will unaided by divinity; his syntactical construction even begins to hint at Hobbes's scandalously demystifying identification, this one not at all authoritarian, of divine grace with innate virtue.[34] By focusing solely on the human acceptance of a general, "large Grace" rather than its particular supernatural dispensation, Milton charts a historical movement from a centralized to a decentralized paradigm of organization, a passage from God's willful exercise of extraordinary providence to man's exercise of free will in the more mundane sphere of ordinary providence. Insofar as Michael is functioning as a Mosaic lawgiver to Adam, his description of the law as "imperfet" or incomplete announces the planned obsolescence of his own historical discourse. The Mosaic law exists to be abrogated, and it is this central fact of its inevitable *Aufhebung* that defines both the legalistic teaching of Moses and the providential history of Michael. Both teachers, this passage suggests, perform a specific function at a certain historical period which will, as man proceeds along the historical trajectory toward the better covenant, undergo an inevitable supersession by a higher mode. In the case of Michael's pedagogy, his lesson of the continual judgment from above must

---

34. "For the proper use of the word *infused*, in speaking of the graces of God, is an abuse of it; for those graces are Vertues, not Bodies to be carryed hither and thither, and to be powred into men as into barrels" (*Leviathan*, p. 441). Danielson, in *Milton's Good God*, pp. 58–91, describes the difference between Calvin's doctrine of selective and "irresistible" grace and Milton's Arminian belief in a universal and "sufficient" grace. But Danielson, I think, underestimates the degree to which the effects of grace in *Paradise Lost* are contingent more on human than on divine will.

necessarily give way to a sense of the individual whose existence as an absolutely free agent depends on the regularity of natural law.

When we consider the sheer excess of Michael's interventionist rhetoric, his assertions of human helplessness amid the unsearchable ways of God, Adam's final lines in the poem are actually quite surprising. His unexpectedly cheerful rejoinder jars with Michael's divine determinism and underscores the revolutionary implications of the organizational tension that reverberates throughout the epic. Adam promises

> ever to observe
> [God's] providence, and on him sole depend,
> Merciful over all his works, with good
> Still overcoming evil, and by small
> Accomplishing great things, by things deem'd weak
> Subverting worldly strong, and worldly wise
> By simply meek.
>
> (12.563–69)

Nowhere is the ideological disparity between teacher and student more conspicuous: whereas Michael has justified the ways of God by deferring ultimate justice to the last days, when God in his glory shall bring "vengeance to the wicked" (12.541), Adam, not at all unlike Gerrard Winstanley, vindicates God here on earth with a progressive hope in the temporal subversion by "things deem'd weak" of things more "worldly strong." We have learned, in our reading of Winstanley, to consider the means by which weakness will overcome strength. By what agency will Adam's revolutionary subversion be effected? Adam insinuates this crucial question of the agent of progress into a pointedly subversive syntactical confusion. His last lines, from one perspective, can be read like Michael's to attribute these triumphant acts of "overcoming," "Accomplishing," and "Subverting" to the grammatical agency of God ("him" in line 564). But the subject of these three optimistic participles can just as easily be Adam's "I" (561), the possible human source of this gradual yet hopeful change. This ambiguity of attribution reflects the historical dialectic of divine and natural law, as if somewhere within these last lines Adam employed the rhetorical device known as anacoluthon, a midsentence shift in grammatical subject that in this case would signify a much more important shift in historical agency. Amid this confusion of subject and action, Adam proves himself to be more "greatly instructed" than we may have at first assumed: his final speech encompasses the entire movement of Michael's universal history, reproducing on the level of grammar the transition from an authoritarian imposition of Mosaic law to a more liberal reliance on rational human nature.[35]

Given the movement Book Twelve has established from shadowy types to truth, from the Mosaic law of sin and punishment to a more vitalist law of nature and its natural processes of justice, the strongly interventionist representation of the actual expulsion which follows might come as a surprise. I want now to examine in some detail the narrator's depiction of the expulsion, the event for which the poem has, surely, overprepared us. The cherubim assisting Michael's expulsion descend from heaven to guard the gates of Paradise:

> and from the other Hill
> To thir fixt Station, all in bright array
> The Cherubim descended; on the ground
> Gliding meteorous, as Ev'ning Mist
> Ris'n from a River o'er the marish glides
> And gathers ground fast at the Laborer's heel,
> Homeward returning.
>
> (12.626–32)

At this exemplary instance of the poem's mythopoesis of extraordinary providence, Milton serenely compares the descent of the cherubim as they prepare for the expulsion to the mist that ascends from a river and follows the homeward journey of a peasant laborer. Although this narrative of angelic descent implicitly affirms God's arbitrary power to intervene in the natural realm, the poem's last extended figure raises the question of causation that can be seen to underlie all of the poem's accounts of the expulsion.

Milton's simile forges a relation between the activity of the mist and the unstated but certainly purposeful action of the cherubim. In order to signal this relation as antithetical, the text opposes the types of motion that characterize the two sides of the simile's equation: the celestial descent of the angels and the autochthonous ascent of the mist. Although the simile voices no explicitly causal link between the gliding mist and the peasant's homeward journey, the urgency behind the image of a mist that "gathers ground fast at the Laborer's heel" suggests it is, perhaps, this mist itself that dogs the laborer's step and pushes him to his appointed end. In making this force a

<hr/>

35. Given the commitment to the liberal organizational discourses Milton conveys in his prose works, it is perhaps not surprising that he concludes his career as a political writer with this same tentative faith in the agency of the passive revolutionary: on the eve of the Restoration in 1659, his world going on "To good malignant, to bad men benign," Milton insists it is still possible for "conscientious men, who in this world are counted weakest," to overcome "the force of this world" (*A Treatise of Civil Power*, 6:22).

vaporous propulsion, a force that seems to border, like Milton's angels, on the limits of the material and the immaterial, Milton surrounds this prod at the laborer's heel with an interpretive penumbra that leaves indeterminate the controlling cause of human action. In its evocation of the "rising Mist" assumed earlier by Satan (9.74–75, 180, 637–38), this evening mist appears to refract the menacing glow of demonic power. But although the image of the mist may invite its metaphorical association with a supernatural force beyond man's control, a rationalistic reading in line with the poem's material narrative also asserts itself.

In his 1651 monistic medical treatise, Francis Glisson attributes a dele-terious physical agency to "vaporous exhalations," tracing some of the causes of a pathology such as rickets to "great Marishes that are obnoxious to much rain and showers"; these marshy mists and vapors, he argues, are able to "drive away and dissipate" the "inherent Spirits of the parts" of the body.[36] It is just this vitalist science of the physically expulsive agency of vaporous exhalations that Milton exploits in this simile, as the evening mist evoked at the end of his epic yields to a literalist identification with those vaporous emanations that have been represented throughout the poem as crucial op-erants in the vitalistic process of Milton's self-contained monistic universe (2.397–402, 5.185–91, 5.423–25, 10.693–95).[37] The simile's engagement of a conflict between ordinary and extraordinary providence, between a liberal and an authoritarian model of agency and organization, is perhaps best cap-tured in Milton's adjective "meteorous." Marvell, in *Upon Appleton House*, had described Mary Fairfax as a "*Comet*" in order to raise the question of the source of her authority: she was to be identified as either a celestial "star new-slain" or a vapor that the "Earth exhale[s]" (684–86). Milton, too, calls on the comet's capacity to crystallize the question of agency. "Meteorous" in this passage names the angels' gliding descent from their supernal origin in the law of God, even as it evokes the early modern belief that meteors originate as vaporous exhalations from the soil.[38]

Just as the epic narrative represents the climactic event of the expulsion as an act of heavenly manipulation, the image of the mist reasserts the vitalist world of ordinary providence, the animistic cosmos in which God exerts his

36. Glisson, *Treatise of the Rickets*, pp. 166, 168.

37. For a general discussion of the period's science of vapors and exhalations, see Marjara, *Contemplation of Created Things*, pp. 177–86.

38. The ambiguity surrounding the origin of meteors is exploited earlier in the poem when Uriel glides through the sky as swiftly as "a shooting Star / In *Autumn* thwarts the night, when vapors fir'd / Impress the Air" (4.556–57). The vitalist doctrine of the autochthonous origin of meteors is stated in Harvey's *De motu cordis* (1628): "For the earth being wet, evaporates by the heat of the Sun, and the vapours being rais'd aloft are condens'd . . . and by this means here are generated, likewise, tempests, and the beginning of meteors" (p. 59).

control over man not by arbitrary judgments but more simply by placing him among those pre-established natural laws that foster liberal justice. But the expulsion narrative continues as the mistlike army of cherubim proceeds in its march through the garden and forces the ejection of Adam and Eve from Paradise:

> High in Front advanc't,
> The brandisht Sword of God before them blaz'd
> Fierce as a Comet; which with torrid heat,
> And vapour as the *Libyan* Air adust,
> Began to parch that temperate Clime; whereat
> In either hand the hast'ning Angel caught
> Our ling'ring Parents, and to th' Eastern Gate
> Led them direct.
>
> (12.632–39)

In this powerful attestation to the strength of Michael's sword, the sword of justice that has on so many occasions violently enforced the will of God (5.316–34), Milton provides us with a final reminder of that alternative vision of historical justice with which he began the story of the expulsion. Michael's sword, "with torrid heat, / And vapour as the *Libyan* Air adust, / Began to parch that temperate Clime," as if this heat were precisely that natural, physical reaction of "those pure immortal Elements" of Paradise that "Eject [man] tainted now, and purge him off / As a distemper" (11.50–53). Like the comparison of the punitive cherubim with the evening mist, the images of vapor and heat that Milton attributes to Michael's sword evoke the vitalist conjunction of matter and spirit expounded so beautifully by Raphael. In opposition to the explicit narrative of divine intervention, these pneumatological images suggest the counterpossibility that in a reality beyond the heuristic tropes of extraordinary providence, the expulsion of Adam and Eve has occurred by means of the mundane interaction of natural elements.

   It is the recognition of this systematic opposition between the attribution of cause to the arbitrary will of God and to the independent, self-regulating elements of nature which best prepares us, I think, to read the poetics of agency behind the last lines of *Paradise Lost*:

> Some natural tears they dropp'd, but wip'd them soon;
> The World was all before them, where to choose
> Thir place of rest, and Providence thir guide:
> They hand in hand with wand'ring steps and slow,
> Through *Eden* took their solitary way.

In assigning to Adam and Eve the continued guidance of "Providence," Milton signals the conflict in the discourses of agency and organization with which this chapter has been concerned: as his *Christian Doctrine* so clearly lays out, the word *providence* inhabits two fundamentally distinct discourses of agency, and Milton's poem does not immediately disclose the status of this final "Providence" as ordinary, extraordinary, or both. The irreconcilability of the meanings of providence we have examined might suggest that the poem ends in a state of radical undecidability, that an insoluble contradiction in organizational philosophy disables the poem's production of a conclusive definition of the first parents' "guide," disabling therefore a conclusive sense of the degree of guidance necessary for the ideal polity. But Milton has struggled in these final books to remove the organizational, and therefore political, antinomies that structure Michael's lesson from the realm of immanent contradiction and to imagine them as elements in a historical transformation. It is Michael's persistent lesson of the historicity of Mosaic teaching, of the tremendous paradigm shift from types to truth, on which Milton depends to determine the nature of the ideologically mysterious "Providence" that oversees the poem's conclusion.

I would argue that in the last books Milton attempts to impose a linear, temporal resolution onto the immanent contradictions in organizational philosophy that fracture the political theology of his entire poem. Michael's historical narrative charts a historical process whereby the Israelites move from dependence on heavenly guidance to a guide that can be located only within themselves (12.485–90). Much like the Israelites, who lose their angelic leader at the Incarnation, at which point they must rely on themselves, Adam and Eve lose their pedagogical guide after the conclusion of Michael's history:

> In either hand the hast'ning Angel caught
> Our ling'ring Parents, and to th' Eastern Gate
> Led them direct, and down the Cliff as fast
> To the subjected Plain; then disappear'd.
> (12.637–40)

The simple predicate, "then disappear'd," signals a momentous shift in the relationship between deity and creature in *Paradise Lost*. This loss of divine agency may assume a poignant, elegiac quality, but the disappearance of Michael, the angelic apotheosis of authoritarian rule, has the effect of situating Adam and Eve solely within the jurisdiction of the law of nature, the law whose institution enabled Milton's construction during the Vitalist Moment of a philosophy of the egalitarian organization of the polity. Just as the

Incarnation in Book Twelve provides the fulcrum for the shift from imposed law to human determination, Michael's disappearance here impels the movement of Adam and Eve from God's authoritarian control to their more solitary placement within the liberally organized realm of a vitalist nature. It is not until this moment, it could be argued, this moment on the "subjected Plain" when Adam and Eve are released from subjection to hieratic divinity, that we have represented before us the birth of the individual with her seemingly autonomous subjectivity.

Like the larger historical movement from the theocentric to the anthropocentric basis of causation, the mysterious term "Providence thir guide" undergoes a local, semantic transition. The naturalistic elements of the narrator's description of the expulsion suggest the new identity of this guide as the inviolable rule of Milton's ordinary providence—that law of nature that may always work toward the moral good of man but that can be distinguished from the direct intervention of an anthropomorphic God. We can see how after Michael's departure, the poem's assertion of "Eternal Providence" shifts away from the appeal to the anthropomorphic God of Scripture and moves toward the liberating faith in the uniform structure of natural and moral law. When subjected to the typological schema Michael himself announces, the extraordinary providence of the almighty God who looms so large in Michael's history can be identified as a "shadowy Type," a heuristic model useful for the proper instruction of the fallen man. And as no accommodated "Type" is sufficient forever in itself, Michael's type of providence is fulfilled in the recognition of the "Truth" of nature's rational operation under the divinely established laws of ordinary providence. Although generations of readers have taken solace in the continued guidance of Adam and Eve by what is assumed to be the controlling force of an extraordinary providence, such interpretive assumptions have failed to register the complex transformations Milton has wrought on providential theology, and on the entire range of politically consequential organizational philosophies that that theology so often subtends. There is a sense in which Milton's stated goal to "assert Eternal Providence" may acquire its ultimate meaning from the Latin root, *asserere*, to declare a slave free. *Paradise Lost* can engender its theology of free will, its politics of self-rule, and its ethos of individualism only by liberating providence from the tyrannical bonds of an authoritarian logic.

6 Margaret Cavendish and the Gendering of the Vitalist Utopia

> Then she desired to be informed whither Adam fled when he was driven out of the Paradise. Out of this world, said they, you are now Empress of, into the world you came from.
>
> —MARGARET CAVENDISH, *The Blazing World* (1666)

I have examined the implications and the complications of the authoritarian takeover in the last two books of Milton's otherwise liberal, vitalist poem. I have not had occasion to mention, however, that Michael's largely authoritarian theology in *Paradise Lost* constitutes a discourse intended solely for the ears of Adam. Eve, for her part, had not been permitted an audience with Michael, who told Adam he had in mind for Eve something less lofty than Adam's visionary ascent to the "top / Of speculation" (12.588–89): "Ascend / This Hill, let *Eve* (for I have drencht her eyes) / Here sleep below . . . / As once thou slep'st, While Shee to life was form'd" (11.366–69). Milton does not inform us of the precise content of the dreams given Eve in her angelically induced state of sleep. But Eve's vague report that her dreams were "propitious" (12.612) invites our speculation on what a female-oriented counterpart to Michael's authoritarian pedagogy might look like. Michael, it is true, suggests to Adam, at the end of Book Twelve, that Eve's oneiric instruction has had the same imperious tenor as the vision offered Adam: the archangel has "with gentle Dreams . . . all her spirits compos'd / To meek submission" (12.595–97). But I would like, for a moment, to ascend my own top of speculation and put forward another, entirely conjectural, theory of the nature of Eve's undisclosed lesson in historical agency.

We have seen throughout this book the curious tendency of the various discourses of vitalist agency to figure the new autonomy of self-moving matter as an abstract principle of autonomous femininity. Van Helmont had championed women, including Eve, as the redemptive bodily force behind his vitalist theodicy; Harvey's vitalism had led to his discreditable thesis in the *De generatione* of the self-sufficient reproductivity of the female ovum; Marvell's alignment of organic autonomy and female virginity had suggested the essentially feminine structure of the passive revolution; and Milton's account of the vitalist emergence of earth brought out a vision of autonomous female generation that burst the normative constraints of his own patriarchal domestic politics. Given what was at its origins the potentially feminist logic of animist materialism, I would venture to suggest that the content of Eve's gentle alternative to the scriptural history Adam receives in Books Eleven and Twelve is a specifically natural history, the liberatory organicist science of self-willing, self-moving matter a brief engagement of which Milton presented us in the form of the vitalist account of expulsion. If Eve is now sleeping as Adam had slept "while Shee to life was form'd," we can reasonably assume that her sleep, too, is productive of a new and unprecedented form of life. The new creature of Eve's dreams is perhaps the radically self-determining individual imaginable only in the context of an implicitly feminist vitalist philosophy.

The subject of this chapter is the natural philosophy of Milton's contemporary, Margaret Cavendish, Duchess of Newcastle. In *The Blazing World*, the romance narrative she appended to her *Observations on Experimental Philosophy*, Cavendish identifies her alternative planet as that same Paradise from which "Adam fled" and which her heroine, one of the figures for the author herself, is "now Empress of."[1] Neglecting in this account any mention of Adam's helpmeet, Cavendish invites our identification of the Empress with Eve, an Eve who in the powerfully utopian space of Cavendish's fictional and scientific writings has been left to govern Paradise alone. Although Milton, in mentioning Eve's dream, remains understandably silent about the theoretical contours of a feminist vitalism, Cavendish engages the science of animist materialism with the unembarrassed intention of exploiting the revolutionary potential of its antipatriarchal logic. We might begin this investigation of the discursive ties that bind Milton and Margaret Cavendish by considering the public appearance of Milton's epic, in 1667, in conjunction with an event occurring in London just a few months earlier, an incident that invited at the time a great deal more attention, and certainly more scandal, than the appearance of *Paradise Lost*. After considerable debate, the men who constituted London's prestigious Royal Society had voted to permit, for the first time in its history, a visit by a woman.[2] On the thirtieth of May, Margaret Cavendish, the Duchess of Newcastle, gained entrance to the Royal Society, the institution for scientific experimentation that was in

1. Margaret Cavendish, *The Blazing World and Other Writings*, ed. Kate Lilley (Harmondsworth: Penguin, 1992), p. 170. All further quotations from *The Blazing World* will be drawn from this edition and cited by page number in the text.

2. See Samuel I. Mintz, "The Duchess of Newcastle's Visit to the Royal Society," *JEGP* 51 (1952): 168–76.

no small way responsible for the complex phenomenon we have come to know as the Scientific Revolution. With an impressive retinue of female attendants, Cavendish was afforded a private exhibition of the period's most important and celebrated experiments. The prominent chemist, Robert Boyle, demonstrated his famous air pump and measured for the duchess the weight in carats of a quantity of air. According to the diarist Samuel Pepys, who was in attendance, Boyle showed her as well an experiment of "colors, another of lodestones, microscopes, and of liquors: among others, of one that did, while she was there, turn a piece of roasted mutton into pure blood, which was very rare." The duchess, Pepys continues, was "full of admiration, all admiration."[3]

The respect accorded her status as the wife of the duke of Newcastle may itself have been sufficient for her admittance to the Royal Society. During the first Civil War, her husband, William Cavendish (then only a marquis), had served Charles I as the foremost general of the king's troops from 1642 to 1644. At the important battles at Tadcaster, Wakefield, and Adwalter Moor, Newcastle had encountered the formidable troops under the generalship of Marvell's patron, Sir Thomas Fairfax. After an ignominious battle at Marston Moor in 1644, Newcastle resigned his commission and fled to Paris, moving from there to Rotterdam and then to Antwerp, where he remained one of the few Royalists not pardoned by Parliament after the regicide in 1649. The young Margaret Lucas had met and married the general during his retired exile on the Continent, and she returned with him to England at the Restoration of Charles II, who made her husband a duke in 1665.

However prominent her husband, though, Margaret Cavendish can surely be said to have garnered in her own right a formidable reputation. She had by 1667 published thirteen volumes of poems, prose romances, essays, letters, and philosophical treatises, not the least of which was her *Observations upon Experimental Philosophy* of the previous year, an attack on the value of telescopes and microscopes, indeed on all the practices of experimental science fostered at the Royal Society. It must be said that one of the prime motives behind Cavendish's remarkable and unprecedented literary productivity was the desire for fame. "That my ambition of extraordinary Fame, is restless, and not ordinary, I cannot deny," Cavendish would later confess.[4] And the success, at least by 1667, of that restless ambition can be measured by the appearance of her name in another major poem of that year, Marvell's "Last Instructions to a

3. *The Diary of Samuel Pepys*, ed. Henry B. Wheatley, 6 vols. (New York: Harcourt, Brace, 1938), 6:324.
4. "Preface," *Natures Picture Drawn by Fancies Pencil to the Life*, 2d ed. (London, 1671), sig. b2ᵛ.

Painter." In a mean-spirited portrait of the Duchess of York, Marvell invokes what he no doubt considered the intellectual pretenses of the wife of Newcastle, Fairfax's bitter enemy: "Paint then again *Her Highnesse* to the life, / Philosopher beyond *Newcastle's* Wife" (49–50). Margaret Cavendish's eccentric dress, her self-consciously vulgar demeanor, and her idiosyncratic displays of intellect had earned her the unhappy sobriquet "Mad Madge of Newcastle." Pepys had expressed concern that the presence of such an outlandish figure at the Royal Society would invite the public's abuse: "the town would be full of ballads of such a visit."[5] He recalls spotting Cavendish earlier that month, driving through London in her "coach, with 100 boys and girls running looking upon her."[6] If such accounts by her contemporaries are any indication, it would not be an exaggeration to suggest that by May 1667 Cavendish had become the object of an obsessive public interest.[7]

The old Puritan revolutionary John Milton, fallen on the evil days of the Stuart Restoration, was not, of course, a member of the Royal Society. And I do not wish to suggest any undue historical significance lurking behind the temporal proximity of Margaret Cavendish's visit and the publication of Milton's epic. But in spite of the considerable gap in academic training, social prominence, and political conviction that separates the sixty-year-old Puritan iconoclast and the forty-four-year-old Royalist duchess, there are some ideologically informed philosophical positions that draw into the same discursive sphere the literary practices of both writers. Cavendish was herself a published poet, and I think there is a good deal of evidence to suggest her familiarity with, and even indebtedness to, Milton's 1645 collection of poems.[8] But the question of a direct literary influence between Milton's work and any of the fifteen original volumes of verse and prose Cavendish published between 1653 and 1668 is not here my subject. Milton and Cavendish

5. *Diary of Samuel Pepys*, 6:324.

6. Ibid., 6:295.

7. See Pepys's entries for March 18, April 11, and April 26, 1667; *Memoirs of the Court of Charles the Second, by Count Grammont* (London, 1846), p. 134; *Letters from Dorothy Osborne to Sir William Temple* (New York, 1888), pp. 111–13; *The Diary of John Evelyn*, ed. Austin Dobson (London, 1908), p. 254, entry for March 18, 1667. The important biographies of Cavendish are Henry Ten Eyck Perry, *The First Duchess of Newcastle and Her Husband as Figures in Literary History* (Boston: Ginn and Co., 1918); Douglas Grant, *Margaret the First: A Biography of Margaret Cavendish* (London: University of Toronto Press, 1957); and Kathleen Jones, *A Glorious Fame: The Life of Margaret Cavendish, Duchess of Newcastle, 1623–1673* (London: Bloomsbury, 1988).

8. Although Isaac Disraeli, in his *Curiosities of Literature* (London, 1833), 2:61, reproduces the eighteenth-century speculation that Milton was indebted to Cavendish's "Dialogue between Melancholy and Mirth" for his "L'Allegro" and "Il Penseroso," Milton's paired poems were probably composed in 1629, when Cavendish was six years old. See Perry, *First Duchess of Newcastle*, pp. 177–78.

are notable participants in the seventeenth-century literary practice of scientific speculation. Drawing their influences, though in various ways, from the philosophies of matter that flourished in the Vitalist Moment, the philosophy especially, I believe, of J. B. van Helmont, both writers find themselves in often untenable literary binds as they struggle to accommodate divergent and contradictory forms of sanctioned truth.

It is their common deployment of a sophisticated philosophy of monism that accounts for the most disquieting points of contact between the idiosyncratic literary projects of Milton and Cavendish. In *Paradise Lost*, Milton endows the earth with a radical capacity for self-perfection by positing a "one first matter all" that extends from the grossest matter of earth to the subtlest spirits of heaven, ascending even further to encompass the essence of God himself (5.469–79). But for the Fall, explains Raphael, the matter of our human bodies might even inch itself up the great chain of being to become part of the rarefied substance of divine spirit. Like Milton and like Hobbes, Cavendish dismisses as oxymoronic the orthodox spiritualist insistence on "immaterial substance." Like the "one first matter all" of Milton's monistic universe, Cavendish's cosmos consists solely of the infinite and indivisible substance she calls "Only Matter."[9] She draws like Milton on the strands of revolutionary political sentiment that had begun, in the Vitalist Moment, to enwrap the doctrine of monism. Having articulated this particular strain of correspondences, I can say now that it is the natural philosophy of Cavendish, and not Milton, Marvell, or Harvey, that constitutes the century's most dramatic attempt to bend the discourse of monistic materialism to a radical cultural end. Margaret Cavendish, I argue, brings her natural philosophy to bear on an ethical project that could only with great difficulty be assimilated to the intended function of Milton's epic. The goal behind the lessons in spiritualized matter that fill so many volumes of her natural philosophy is to supply the metaphysical foundations for a social agenda for which she had almost no contemporary support—the liberation of women from the constraints of patriarchy. As we will see, however, Cavendish's science, and the liberal feminism it subtends, falls prey to the same pressures of contradiction that beset all the theorists of a liberal vitalism.

The Science of Strength and Weakness

Of the sixteen volumes of prose and verse Cavendish published between 1653 and 1668, what has commanded the most interest today are not the

9. Margaret Cavendish, *Philosophical and Physical Opinions*, 2d ed. (London, 1663), p. 1.

scientific treatises on the nature of matter but the many epistles and prefaces she appended to her books. It is in the personal, informal setting of the preface that Cavendish fashions a discourse whose outlines have been justly identified as "feminist."[10] She evinces a protosociological sense that female intellectual behavior is more attributable to society's prejudicial sexual hierarchy than to immutable female physiology. In an epistle, for example, addressed "To the Two Universities," Cavendish exhorts the faculties of Cambridge and Oxford to encourage the formal education of women, "lest in time we should grow irrational idiots . . . for we are kept like birds in cages to hop up and down in our houses . . . ; thus wanting the experiences of nature, we must needs want the understanding and knowledge and so consequently prudence, and invention of men."[11] "Our Sex," she writes elsewhere, is "not suffer'd to be instructed in Schools and Universities," the consequence of which exclusion is that "many of our Sex may have as much wit, and be capable of Learning as well as men; but since they want Instructions, it is not possible they should attain to it."[12] The identifiable differences in the intellectual deportment of men and women are often ascribed, as in these passages, not to nature but to society's artificially inequitable system of liberties and constraints. That system operates for Cavendish through a careful segregation of the male and female spheres of motion and action: it is, more specifically, the compulsory confinement of female movement and action to the home that constitutes the primary impediment to women's intellectual progress.

Although the rhetoric of emancipation and equality informs a number of Cavendish's pronouncements on sexual difference, the most common gestures in her prose are assertions in a considerably different ideological key. Cavendish seems committed as well to the notion that women's behavior and thought are uniquely and naturally "feminine" and sometimes even inferior to the behavior and thought of men. In a preface to *The Worlds Olio* (1655), she writes of the "great difference betwixt the Masculine Brain and the Feminine": Nature had fashioned "Mans Brain more clear to understand

---

10. Considerations of Cavendish as feminist include Virginia Woolf, *The Common Reader* (New York: Harcourt, Brace, 1948 [1925]), pp. 98–109; Mary Ann McGuire, "Margaret Cavendish, Duchess of Newcastle, on the Nature and Status of Women," *International Journal of Women's Studies* 1 (1978): 193–206; Sara Heller Mendelson, *The Mental World of Stuart Women: Three Studies* (Amherst: University of Massachusetts Press, 1987); Londa Schiebinger, *The Mind Has No Sex? Women in the Origins of Modern Science* (Cambridge: Harvard University Press, 1989).

11. Cavendish, "To the Two Universities," prefixed to the 1655 edition of *Philosophical and Physical Opinions*, n.p.; quoted in Perry, *First Duchess of Newcastle*, p. 187.

12. Cavendish, Preface to *Observations upon Experimental Philosophy*, n.p.

and contrive than Womans. . . . Women can never have so strong Judgement nor clear Understanding nor so perfect Rhetorick."[13] She makes no attempt to distinguish a social analysis of feminine behavior from an essentializing biological view with which it would seem to be at odds; she forwards vigorous and moving arguments for women's access to the higher spheres of learning, but these arguments are coupled, with little apparent irony, to protestations of women's intellectual unfittedness for those same social institutions. Given the extent of Cavendish's divergent expressions of the status of women, it is little wonder her writing has proved such a source of frustration for historians of early modern English feminism.[14]

We have examined throughout this book the pressure that the organizational imperative could exert on the period's natural philosophical speculation. Surely a primary appeal of natural philosophy was its function as a discursive forum in which ideas of human relations, including the relations between the sexes, could be debated, questioned, and affirmed. Like the texts of Harvey and Milton, Cavendish's scientific treatises embrace a range of conflicting philosophies of material organization that point to irresolvably conflicting models of social and political organization. Her theorization of the role of women, on the discursive level of explicit social commentary, will no doubt always trouble readers with its contradictions and reversals. But I argue here that Cavendish's fascinating, confusing analysis of sexual politics is submitted to a sophisticated reworking in the pages of her natural philosophy. It is in respect to the social significance so often attending hypotheses of material organization that I want to examine the startling scientific conversion Cavendish undergoes in 1663.

Cavendish begins her pursuit of natural philosophy not as a vitalist but a mechanist. Her first work, *Poems, and Fancies*, published in 1653, contains hundreds of lyrics, many of them versified disquisitions on the day's most pressing concerns of natural philosophy. In a characteristic poem, "What Atoms make Fire to burne, and what Flame," Cavendish envisions the mechanics of fire as the violent interaction of differently shaped atoms:

> Thus *Flame* is not so hot as *Burning Coale*;
> The *Atomes* are too weake, to take fast hold.
> The *sharpest* into firmest *Bodies* flye,

13. Cavendish, *The Worlds Olio* (London, 1655), sigs. A4^{r-v}.

14. See, for example, Hilda Smith, *Reason's Disciples: Seventeenth-Century English Feminists* (Urbana: University of Illinois Press, 1982), p. 93; and Marilyn Williamson, *Raising their Voices: British Women Writers, 1650-1750* (Detroit: Wayne State University Press, 1990), p. 38. I am grateful to Jay Stevenson for bringing these works to my attention.

> But if their Strength be small, they quickly dye.
> Or if their *Number* be not great, but small;
> The *Blunter Atomes* beate and quench out all.[15]

Cavendish was perhaps justified in explaining, years later, that this first attempt at natural philosophy, "by reason it is in Verse . . . is not so Clearly or Solidly Expressed, as I might have done it in Prose."[16] Bracketing the possible aesthetic lapses of her verse, we can nonetheless trace some of the dominant philosophical outlines of her earliest scientific speculations. Despite the fact that Cavendish is writing at the end of the Vitalist Moment, the materialist system informing the lyrics in her *Poems, and Fancies* is a decidedly nonvitalist atomism. Like the chaos Milton describes at the end of Book Two of *Paradise Lost*, Cavendish's earliest representations of matter fall in line with the corpuscular theories of matter advanced by the mechanistic philosophers Gassendi, Descartes, and Hobbes, men with whom the duchess had made contact through her husband in Paris.[17] She maintains without hesitation a mechanistic materialism that admits nothing but the forced impulsion of bodies as either cause or effect of physical motion.

As her distinction, in the poem above, between sharper and blunter atoms makes clear, Cavendish was like Milton given to make use, perhaps for its poetic interest, of a superannuated philosophy that grouped atoms by what were believed to be their round, sharp, blunt, and flat shapes. But despite these traces of an outmoded Epicureanism, the dynamics of her materialism is fully consonant with Hobbes's famous declaration that "there can be no cause of motion, except by a body contiguous and moved."[18] The motions of fire, for example, are attributable to the violent interaction of contiguous bodies: "Sometimes for anger, the sparks do flye about; / Or want of room,

---

15. Cavendish, *Poems, and Fancies: Written by the Right Honourable, The Lady Newcastle* (London, 1653), pp. 13–14.

16. Cavendish, *Philosophical and Physical Opinions*, sigs. c2ᵛ–c3ʳ.

17. The contemporary biographer John Aubrey writes: "I have heard Mr. Edmund Waller say that W. Lord Marquis of Newcastle was a great patron to Dr. Gassendi and M. Des Cartes, as well as to Mr. Hobbes, and that he had dined with them all three at the marquis's table at Paris" (*Letters*, ii. 602). Newcastle, we can assume, had a special interest in the scientific pursuits of these three philosophers. Indeed, according to William T. Lynch, "it appears the first evidence we have of Hobbes' interest in either mechanics or optics (apart from the 'Little Treatise') is from Hobbes' letter to Newcastle of 26 January 1634, where Hobbes relates his difficulty in obtaining Galileo's *Dialogues*, as he had promised" (Lynch, "Politics in Hobbes' Mechanics: The Social as Enabling," *Studies in History and Philosophy of Science* 22 [1991]: 306). See Hobbes's letter to the earl of Newcastle, October 10, 1636, in Historical Manuscripts Commission, *13th Report Part II, Manuscripts of the Duke of Portland* (1893), p. 130.

18. *Elements of Philosophy. The First Section, Concerning Body*, in *English Works of Hobbes*, 1:124.

the weakest are thrust out." In these early speculations, motion is simply not possible but for such acts of bodily displacement, the "thrusting out" of the weakest atoms by the strongest. For Hobbes, even the smallest unit of matter seemed to pursue a course of perpetual conquest and self-aggrandizement, since a body's motion was nothing more than its "continual relinquishing of one place, and acquiring of another."[19] As Hobbes's own work made strikingly clear, this principle of the violent displacement that necessarily accompanied bodily motion was easily transferred to the world of human interaction.

Cavendish shares with all the writers we have examined a drive to marshal images from natural philosophy as an organizational foundation for her beliefs about human society and, more specifically, about the interaction of men and women. The battle of the sexes could be waged, discursively, in the debates over the physical constitution of the natural world: the question of the relation of men to women, for example, seems to have informed the widespread speculation about the relation of spirit to matter or of the sun to the moon: both Milton and Cavendish enlist the analogy of the moon's subordination to the sun to evince the rightful dependence of woman on man.[20] Natural philosophers had traditionally taken for granted the gendered constitution of the cosmos—the belief in what Milton calls in *Paradise Lost* those "two great Sexes [that] animate the World" (8.151). Cavendish would not have wanted to quibble with the principle of gendering itself; she consistently portrays Nature as an ingenious though dutiful housewife: "Nature, being a wise and provident Lady, governs her parts very wisely, methodically and orderly; also she is very industrious, and hates to be idle, which makes her imploy her time as a good Huswife doth, in Brewing, Baking, Churning, Spinning, Sowing, &c."[21] Capitalizing on the household autonomy accorded the wise and provident housewife, Cavendish is clearly comfortable here with the identification of nature as feminine: Nature may be a housewife bound by duty to an implicitly male superior, but within the sphere of the household her wise, methodical, orderly government is her own.

If the traditional natural philosophy of the two great sexes demanded a degree of rhetorical respect for the feminine principles of the cosmos, the mechanistic advances of the Scientific Revolution had begun to render such advantageous gender identifications increasingly insupportable. As we have observed in Cavendish's own early poems, the mechanization of the world

---

19. Ibid., 1:109.

20. See *Paradise Lost*, 8:149–50; and Cavendish's *Worlds Olio*, sig. A4ᵛ. The best discussion of the gendering of nature in *Paradise Lost* is Guillory, "From the Superfluous to the Supernumerary," pp. 68–88.

21. Cavendish, *Observations upon Experimental Philosophy*, pp. 101–2.

picture in the mid-seventeenth century seemed to reduce what could be seen as the equitable division of labor between nature's two great sexes to the interplay merely of strength and weakness. What had been the idea of the gendered but complementary forces animating the world was quickly being supplanted by the image of the simple dominance of the strong elements of nature over the weak. For the new chemist or physicist, as Carolyn Merchant and Evelyn Fox Keller have argued, change in the natural world was often little more than the effect of an implicitly masculine force of motion on the passive, even lifeless, elements of an implicitly feminine matter.[22] If the traditional natural philosophy of Mother Nature and God the Father figured sexual hierarchy in the language of relatively reciprocal relations, the new mechanistic science of agency and organization flattened the relation of reciprocity to one of absolute subjection; it rendered natural and inevitable the rights and privileges of physical force. Any organizational analogy between the mechanistic microworld of material particles and the social world of men and women could only accord significance to that sex in possession of greater physical strength.

It is just this problem of the mechanist's reduction of agency to a simple matter of bodily collision that I wish to isolate as the primary precipitant of Margaret Cavendish's conversion from mechanism to vitalism. The implicitly masculine bias of the new materialism presented Cavendish, as both feminist and mechanist, with a considerable dilemma. The sign of this dilemma, I propose, is the unrelenting attention, in her early exercises in cultural commentary as well as in her plays, to the problem of physical strength. In *The Worlds Olio*, Cavendish suggests that the factor of muscular might alone is sufficient to distinguish men and women: "There is a great difference betwixt . . . Masculine Strength and the Feminine . . . for Nature hath made Mans Body more able to endure Labour. . . . This is the reason we are not so active in Exercise, nor able to endure Hard Labour, nor far Travells, nor to bear Weighty Burthens, to run long Jornies, and many the like Actions which we by Nature are not made fit for."[23] Because the capacity to "bear Weighty Burthens" might, at least according to the force-oriented logic of mechanism, be a sufficient signifier of sexual superi-

22. See Merchant, *Death of Nature*; and Keller, *Reflections on Gender and Science*, pp. 33–65. The new physics of active force and passive matter simply recast in a new language the gendered hierarchy structuring an earlier period's articulation of sexual reproduction, a fair example of which is this sentence from the sixteenth-century physician Ambroise Paré: "the seede of the male being cast and received into the wombe, is accounted the principall and efficient cause, but the seede of the female is reputed the subjacent matter, or the matter whereon it worketh" (*Workes of that Famous Chirurgian Ambrose Parey*, p. 885).

23. Cavendish, *Worlds Olio*, sigs. A4ʳ–A5ʳ.

ority, Cavendish is compelled to argue at times that the difference in male and female musculature has its basis not in physiology but in culture. If women were only granted access to the world of masculine sportsmanship, she proposes, they could exercise their limbs to match the physical strength of men's: "let us Hawk, Hunt, Race, and do like Exercises as Men have . . . we should Imitate Men, so will our Bodies and Minds appear more Masculine, and our Power will Increase by our Actions."[24] This attribution of all differences in physical strength to factors of socialization is difficult, however, even for Cavendish to maintain. We find her more typically pressed to acknowledge the limits of this particular application of cultural analysis: "It is true, Education and Custom may adde something to harden us, yet never make us so strong as the strongest of Men, whose Sinnews are tuffer, and Bones stronger, and Joints closer, Flesh firmer than ours are, as all Ages have shewn, and Times have produced."[25]

Cavendish's appeal to size and strength as the ineradicable markers of sexual difference exposes an almost irremediable inscription within the contemporary scientific discourses of agency and force. Since female identity is designated by an absence of physical strength, the female liberation for which Cavendish so often calls is likewise figured in physical terms, as an enlarged field of action or range of motion. This specific identification of human freedom with free physical motion has for Cavendish its important origin in the negative conception of freedom formulated most recently by Hobbes. "Liberty, or freedome," Hobbes wrote in *Leviathan*, "signifieth (properly) the absence of Opposition; (by Opposition, I mean externall Impediments of motion;) and may be applyed no lesse to Irrationall, and Inanimate creatures, than to Rationall."[26] As C. B. Macpherson has explained, the organizational force of Hobbes's philosophy relied on an assumption of the logical relation of physiological motion to the social motion of human beings.[27] As a simple freedom from external and, always implicitly, physical constraints, Hobbesian liberty was reducible to an individual's physical capacity to *move* freely. An impingement on freedom, then, was in essence no more than a physical obstacle, the impeding pressure, most usually, of a stronger body on a weaker. It is with precisely this physicalized conception of personal freedom that Cavendish begins to represent a system of sexual subjection as a system of impediments to physical progress or continued

24. Cavendish, *Orations of Divers Sorts* (London, 1662), p. 225. See the similar statement from Cavendish's *Bell in Campo*, quoted in Jones, *Glorious Fame*, p. 132.

25. *Worlds Olio*, sig. A5ʳ.

26. Hobbes, *Leviathan*, p. 261.

27. C. B. Macpherson, *The Political Theory of Possessive Individualism: Hobbes to Locke* (Oxford: Clarendon, 1962), pp. 76–77.

motion, the institutional cage of the home in which women are kept "to hop up and down."

Although Cavendish relies on a Hobbesian notion of freedom in her appeals for women's liberty, any acceptance of Hobbes's cynical premise that physical force determines social privilege brings with it a disconcerting pessimism about the possibilities of female emancipation. It is difficult to imagine anything but the most patriarchal conclusion derivable from Hobbes's ruthless, scientist view of the priority of physical strength. Mechanism provided masculine dominance with a powerful organizational sanction, and I suspect that it was precisely the untenable nature of such conclusions that impelled Cavendish to distance herself from the mechanical explanation of natural change and the negative conception of liberty it seemed logically to imply. In 1663, Cavendish mounts a surprising, thoroughgoing critique of her own mechanistic principles. At the very moment in Restoration England when the mechanistic theories of atomic matter in motion forwarded by the Royal Society are gaining increasing intellectual assent, Margaret Cavendish repudiates the tenets of mechanism to embrace the animist materialism that had flourished a decade before at the Vitalist Moment. Whereas her acquaintance and sometime translator, Walter Charleton, like the chemist Robert Boyle, had firmly repudiated the Paracelsian and Helmontian vitalism he had embraced as a youth in the early 1650s, Cavendish, like Milton, turns back to an outdated, discredited vitalism with which to critique the authoritarian mechanism of Hobbes and Descartes.[28] Characterizing in 1663 the mechanistic theory of matter as a hypothetical world marked by an unpleasant exhibition of masculine prowess, she rejects the "Opinion of some Wise and Learned Men . . . that all Exterior Motions, or Local Actions or Accidents proceed from one Motion Pressing upon another, and so one thing Driving and Shoving Another to get each other's Place."[29] She insists, on the contrary, that the "actions of nature are not forced by one part, driving, pressing, or shoving another, as a man doth a wheel-barrow, or a whip a horse; nor by reactions, as if men were at foot-ball or cuffs."[30] This implicitly masculinist explanation of motion as a reactivity to violent impulsion is no longer tenable, because motion is the consequence not of reaction, she ar-

---

28. For the defection of Boyle and Charleton from the vitalist camp, see Rattansi, "Paracelsus and the Puritan Revolution," p. 32; and Gelbart, "Walter Charleton." Boyle's *Sceptical Chymist* (London, 1661) charts the debate between mechanism and midcentury Paracelso-Helmontianism.

29. Cavendish, "Another Epistle to the Reader," *Philosophical and Physical Opinions*, sig. c3^v. The best account of Cavendish's philosophical conversion is Lisa Sarasohn, "A Science Turned Upside Down: Feminism and the Natural Philosophy of Margaret Cavendish," *Huntington Library Quarterly* 47 (1984): 289–307.

30. Cavendish, *Philosophical Letters*, p. 95.

gues, but of volition: material bodies can move, according to the newly vitalist Cavendish in 1663, only if they choose to move.

This belief in the volitional power of motion intrinsic to all physical matter becomes the unrelenting, endlessly repeated thesis of all Cavendish's later philosophical texts. In dozens of passages throughout the later scientific treatises, she reiterates these dicta: "Nature moveth not by force, but freely" and "self-moving matter is the only cause and principle of all natural motion."[31] In a world in which all natural matter is alive and self-moving, one need not account for motion in terms of the pressure of one particle of matter on another particle moving with lesser force: the parts of matter do not "drive or press upon each other, for those are forced and constraint actions, when as natural self-motions are free and voluntary."[32] Like blood for Harvey in 1649 and human tissue for Glisson in 1650, matter for Cavendish in 1663 possesses attributes of motional self-determination hitherto reserved for thinking, soulful human beings: "all things, and therefore outward objects as well as sensitive organs, have both Sense and Reason."[33] Bodies of matter, however trivial or inert they appear to the naked eye, all function as "sensitive organs," in possession not only of sense and reason but of perception as well.[34] Nature is not only autonomous; it is as perceptive and conscious and purposive as freely willing men and women. The philosopher, in consequence, who holds this view of self-moving matter frees herself from a resignation to the physics—and the corollary masculinist ethics— of the rule of force.

The singular daring of her particular version of the monist heresy, asserted considerably after the vitalist movement had flourished and faded, was not lost on the duchess. Cavendish no doubt felt obliged to articulate a rigorous justification of her decidedly counterintuitive philosophy of self-moving matter. But the theory she devised to justify this faith in matter's sense, reason, and perception was of such baroque complexity that her learned contemporaries were forced to dismiss it as ridiculous. The Cambridge Platonist Ralph Cudworth took issue, as we will see, with her philosophy's impiety, and the English Cartesian Joseph Glanvill confuted Cavendish's healthy skepticism about the existence of ghosts.[35] But for these few exceptions, her

31. See, for example, Cavendish, *Philosophical Letters*, p. 23; *Observations upon Experimental Philosophy*, sig. g1^v.

32. Cavendish, *Observations upon Experimental Philosophy*, p. 138.

33. Cavendish, *Philosophical Letters*, p. 18. Cavendish evinces her debt to Harvey's late account of a liberal circulation in *The Blazing World*, as the scientists summoned by the Empress reason that "the circulation of the blood" occurs by means of "an interior motion" (p. 146).

34. Cavendish, *Observations upon Experimental Philosophy*, sig. g2^r.

35. See Glanvill's correspondence in *A Collection of Letters, Poems, etc. written to . . . the Duke and Duchess of Newcastle* (London, 1678).

exercises in scientific speculation lay so far outside the acceptable contemporary parameters of natural philosophical explanation that, like Harvey's arguments for reproductive self-sufficiency in *De generatione animalium*, they invited little serious debate. On reading Cavendish's attacks on her philosophical contemporaries in *Philosophical Letters*, Henry More wrote to his learned correspondent Lady Anne Conway that he believed that Cavendish "may be secure from any one giving her the trouble of a reply."[36] Modern readers have left Cavendishean philosophy similarly secure from investigation, heeding Virginia Woolf's caveat that Cavendish's science represents a lamentable attempt by a woman to erect with the tools of masculine learning a "philosophic system that was to oust all others."[37] However far afield this philosophic system from the period's more orthodox endeavors in scientific speculation, Cavendish's science merits our attention for its unique and powerful engagement of the urgent problems of agency and organization with which her culture was so thoroughly absorbed. I propose that we examine how the duchess's revised model of physical agency provided an important conceptual model for her interest in feminine agency, and how her philosophy of organization provided a language in which to explore an oppressively gendered society and the place in that society of an individual both free and female.

The Commonwealth of Matter

We have seen how Cavendish's vitalism works to replace compulsion with choice in the physics of material bodies. This theoretical substitution, however, runs counter to the perception of colliding objects in the empirical world of post-Galilean physics, the world in which force and not free will seemed so clearly to provide the dominant impetus for motion. It is the challenge of Cavendish's physics to controvert the empirical faith in the primacy of external force by directing the natural philosophical gaze away from the interaction of material bodies in observable space. Her theory must focus instead on the much subtler interaction of the "parts" of matter that dwell *within* each material body. Motion, she claims, has little to do with external force or impulsion. The motion of each and every material body, whether

36. *Conway Letters: The Correspondence of Anne, Viscountess Conway, Henry More, and Their Friends, 1642–1684*, ed. Marjorie Hope Nicolson (New Haven: Yale University Press, 1930), p. 237.

37. Woolf, *The Common Reader*, p. 109.

animal, vegetable, or mineral, is for Cavendish the consequence of the organization and reorganization of its three distinct types of internal parts: the "rational" animate matter, the "sensitive" animate matter, and, finally, the inanimate matter. Rejecting an unmodified vitalist faith in a universally animated cosmos,[38] Cavendish permits herself to follow Descartes and Hobbes (not to mention Milton's Raphael, in his insertion into chaos of the "black tartareous cold Infernal dregs"), admitting to her vitalist materialism the category of inanimate matter, which must serve its function as the raw material substrate on which the animate matter works. It is the self-moving animate parts of matter that bear the responsibility for orchestrating the movements and reconfigurations of inanimate matter. How precisely the animate parts of matter fulfill this duty, how these free and self-determining elements can fashion in all its order and variety the phenomenal world, is the central focus of Cavendish's science.

The first and most rarefied components of any material body are the "rational" parts, which are in possession of precisely those determinants of reason found in individual human beings. The rational parts of matter use their power of self-understanding, knowledge, and even perception to move themselves and command a knowledge of everything around them. For all this wisdom, however, these rational corpuscles lack the actual physical strength to move and reconfigure the inert, un-self-moving parts of inanimate substance that everywhere surrounds them. They are dependent for this task on the less rarefied "sensitive" parts of animate substance, parts that lack some of the intellectual and perceptive capacity of "rational" matter but possess a muscular grip sufficient to hoist and carry the weighty "burthens of other parts" of inanimate matter.[39]

Because all material bodies, even those seemingly inanimate bodies referred to by Cavendish as "outward objects," possess the full complement of the rational, sensitive, and inanimate parts of matter, no observable motion, not the slightest shudder, can occur without the interaction of these three mutually dependent classes. Within the framework of the duchess's theory, for example, we arrive at a new understanding of that type of motion that had proved so paradigmatic for the empirically minded mechanists, the movement of a billiard ball. For Cavendish, a billiard ball initially at rest is set in motion not because another ball, impelled with force, has struck it. Such an appeal to the causal priority of external pressure would be a lapse into the misguided explanation of the mechanists. Cavendish's billiard ball is

38. Cavendish, *Observations upon Experimental Philosophy*, sig. h1[h]ʳ–h1[h]ᵛ. See below, pp. 222–23.

39. Cavendish, *Observations upon Experimental Philosophy*, p. 192.

seen to move because the other billiard ball has *occasioned* its motion, but this act of occasioning can in no way constitute a direct cause of its motion.[40] All material objects are self-moving, the cause of any object's motion lying necessarily within that body itself. The struck billiard ball moves because its rational particles of matter have actually perceived the hasty approach of the rival parts of matter in the other billiard ball. How else, asks Cavendish, "shall Parts work and act, without having some knowledge and perception of each other?"[41] On exercising their powers of perception and reason, the first ball's rational particles assess the effects of the impending collision with the second ball and choose, freely, to move. Because these rational parts of matter are themselves too weak to organize the inanimate dregs around them, they must command their sensitive brethren to drag bodily the remaining, inanimate, particles that make up the ball. The mainspring of change behind the complex action we see as the ball's movement—the sufficient cause, in fact, by which all bodies must move—is the voluntary obedience of the sensitive parts of matter to their rational superiors. All changes in the physical world, and, therefore, all conceivable changes, involve this tripartite process: the initial exercise of the infinite wisdom and perceptive powers of the rational matter, the demands of this rational matter on the laboring sensitive matter, and, finally, the free consent of the sensitive matter to obey those demands.

The image on which Cavendish most frequently draws to illustrate this extraordinary theory of motion is the construction of a house:

> as in the Exstruction of a house there is first required an Architect or Surveigher, who orders and designs the building, and puts the Labourers to work; next the Labourers or Workmen themselves, and lastly the Materials of which the House is built: so the Rational part, in the framing of Natural Effects, is, as it were, the Surveigher or Architect; the Sensitive, the labouring or working part, and the Inanimate, the materials, and all these degrees are necessarily required in every composed action of Nature.[42]

Cavendish's two classes of animate matter enjoy a perfect and unconflicted division of labor, the rational parts assuming the responsibility for designing and framing the laborious actions of the sensitive parts on the remaining unself-moving elements of inanimate matter. The motion that results from this labor, hardly the lifeless reaction to external pressure imagined by the mech-

---

40. The distinction between occasion and cause is addressed in *Observations upon the Opinions of Some Ancient Philosophers*, p. 3, bound with *Observations upon Experimental Philosophy*.

41. Cavendish, *Observations upon Experimental Philosophy*, sig. f1ᵛ.

42. Ibid., sig. h2ʳ.

anists, is an elaborately designed and carefully constructed architectural product. Far from the simple accounting of the temporal movement from cause to effect, Cavendishean motion depends on a spatialized image of an entire social organization. Every outward object is a micro-utopia, and every physical motion is the result of an intricate and spontaneous set of harmonious micropolitical negotiations.

It is important, I think, to understand the attraction for Cavendish of this theory of an independently ordered society of the politically active parts of matter. One of the primary appeals lies surely in the "liberal" manner in which it can account for those unpredictable historical occurrences Milton might have felt obliged to impute to "extraordinary providence." It is with a considerable explanatory elegance that Cavendish can justify the ways of matter, accounting for those actions in nature that even the scientists at the Royal Society were compelled to explain by resorting to a providential deus ex machina. Robert Boyle, for example, never fully forsook the model of the centralized, voluntarist organization of the universe: "seeming anomalies" in the otherwise smooth flow of natural process were for Boyle, as for Milton's archangel Michael, the action of a God who "sometimes purposely overrules" the identifiable patterns of corpuscular motion. For Boyle, as for Michael, it is for reasons of moral pedagogy that God interrupts the course of ordinary providence in order "to execute his justice; and therefore plagues, earthquakes, inundations, and the like destructive calamities, though they are sometimes irregularities in nature, yet for that very reason they are designed by providence, which intends, by them, to deprive wicked men of that life, or of those blessings of life, whereof their sins have rendered them unworthy."[43]

The anomalies of nature that for Boyle could still be explained as the result of the extraordinary intervention of an often punitive deity are for Cavendish the result of the local disagreements and acts of dissent among the independently configured parts of matter: "for although several parts are united in one body, yet are they not always bound to agree in one action; nor can it be otherwise; for were there no disagreement between them, there would be no irregularities."[44] In her earliest formulations of self-moving matter, Cavendish had ascribed the cause of inexplicable natural phenomena to

43. *Works of Boyle*, 5:164, 198; quoted in J. R. Jacob, *Boyle and the English Revolution*, pp. 160–61. See John Henry, "Henry More versus Robert Boyle: The Spirit of Nature and the Nature of Providence," in *Henry More (1614–1687) Tercentenary Studies*, ed. Sarah Hutton (Dordrecht: Kluwer, 1990), p. 61. Henry notes elsewhere that Boyle and other mechanistic philosophers also questioned this faith in a voluntarist God. See his "Occult Qualities," p. 354.

44. Cavendish, *Observations upon Experimental Philosophy*, p. 168.

failed acts of perception and inadvertent slips in communication among the
various parts of matter, since "self-moving matter," she explained, "may
sometimes erre and move irregularly."[45] By 1666, however, in *Observations
upon Experimental Philosophy*, she pronounces the parts of rational matter
infallible: "they cannot readily err, unless it be out of wilfulness to oppose
or cross each other."[46] Sufficient to stand though free to fall, her rational
parts of matter behave in a disorderly fashion not out of corruption but out
of choice. Cavendish, of course, takes pains to write "Pure Natural Philosophy, without any Mixture of Theology."[47] But it is nonetheless her vision
of an essentially *unfallen* world of free and rational agents that enables her
to maintain the closed economy of motion and change that, as we have
observed, was so central to the liberal mechanics of Paradise in Milton.

We have seen the reactions of the orthodox to the position, espoused by
Milton, among others, that the natural world enjoys a certain freedom from
the meddlesome actions of a supervisory God. Cavendish's view, that nature
is not only autonomous but fully conscious, arouses a response of even
greater vehemence. Of all the forms of philosophical materialism attacked
by the Cambridge Platonist Ralph Cudworth, no doctrine invites his impassioned invective quite like the theory of motion advanced by Cavendish and
her vitalist precursors. Although the duchess, like the other philosophers
under Cudworth's attack, is never mentioned by name, it is very likely her
peculiar philosophy of material self-motion that elicits his denouncement, in
his massive *True Intellectual System of the Universe*, of "hylozoic atheism":
the hylozoists, he insists, "conceive grosly both of *Life* and *Understanding*,
spreading them all over upon Matter, just as Butter is spread upon Bread,
or Plaster upon a Wall."[48] This gross materialization of life and understanding
is not only bad chemistry; much worse, Cudworth suggests, it poses a serious
threat to the belief in the omnipotence of God: "*Hylozoick Atheism*, is nothing but the *Breaking* and *Crumbling* of the *Simple Deity*, One Perfect Understanding Being, into *Matter*, and all the several *Atoms* of it."[49] Cavendish

45. Cavendish, *Philosophical Letters*, p. 152. See also *Philosophical and Physical Opinions*,
pp. 43–44.

46. Cavendish, *Observations upon Experimental Philosophy*, p. 192.

47. Cavendish, Preface to *Philosophical and Physical Opinions*, sig. b2ᵛ.

48. Cudworth, *True Intellectual System*, p. 174. Although John Henry has stated, in "Medicine and Pneumatology," p. 28, that "it is perfectly clear" that Cudworth "had no one else in
mind but . . . Francis Glisson," I would insist that Cavendish, whose many treatises of hylozoist
philosophy predate Glisson's *De natura substantiae energetica* (London, 1672), is one of the
primary objects of Cudworth's attack. That Henry has overlooked Cavendish is clear from his
claim that "Glisson took the concept of inherently active matter further than any of his contemporaries" (p. 29).

49. Cudworth, *True Intellectual System*, p. 871.

is careful to insulate herself from such just accusations of impiety, insisting, much like Milton, that natural self-sufficiency is not logically inconsistent with divine authority: "Although Matter is self moving . . . I do not say, that Nature has her self-moving power of her self, or by chance, but that it comes from God the Author of Nature."[50] She goes to some length to establish the rhetorical appearance of orthodoxy, insisting that she will ever "follow the Instruction of our blessed Church."[51] But the logic of self-sufficiency at the heart of her almost exclusively secular natural philosophy brings her nonetheless as close as it was possible to come to an explicit assertion of the decentralized, godless universe of liberal science. Her philosophy, as Cudworth knew very well, was one "so Outragiously Wild, as that very few men could have Atheistick Faith enough to swallow it down and digest it."[52]

The inner workings of Cavendish's physics are, to be sure, distinctly hierarchical; nature is governed by the innately aristocratic rational parts of matter. But the groupings of rational, sensitive, and inanimate matter have been dispersed quite uniformly—one could even say democratically—through the fabric of the physical world. Her material system is so decentralized that Cavendish refuses to privilege the matter of human bodies over the matter of outward objects, or even the matter of the human brain over the rest of the human body: "neither doth this sensitive and rational matter remain or act in one place of the Brain, but in every part thereof; and not onely in every part of the Brain, but in every part of the Body; nay, not onely in every part of a Mans Body, but in every part of Nature."[53] In his final work, the *Treatise on the Energetic Nature of Substance*, Francis Glisson would extend the logic of his monism to attribute to all the tissues of the human body a capacity not only for sense and reason but also for muscular motion, perception, and even desire.[54] And though Milton was not willing to extend quite this degree of sentience to human flesh, he was happy to attribute to the entire angelic body (a body indivisible for Milton into distinct organs and faculties) a tissue-based ability to see, hear, think, and feel; of the bodies of his fellow angels Raphael tells Adam: "All Heart they live, all Head, all Eye, all Ear, / All Intellect, all Sense" (6.350–51). Cavendish, I would argue, stretched her monism further than either Glisson or Milton. It is "not onely in every part of a Mans Body, but in every part of Nature"

50. Cavendish, *Further Observations*, p. 44, bound with *Observations upon Experimental Philosophy*.

51. Ibid., p. 39.

52. Cudworth, *True Intellectual System*, p. 145.

53. Cavendish, *Philosophical Letters*, p. 185.

54. Glisson, *Tractatus de natura substantiae energetica*.

that she locates the highly organized exercise of sense, reason, and perception.

We have examined the dialectic of liberal and authoritarian principles that structures the works of Marvell and Milton. Both poets voice a potentially dissenting materialism, articulating a form of hylozoist materialism to combat an ideologically unsatisfactory orthodox providentialism. Although they imagine the possibility of a liberal humanist ethics logically contingent on the autonomous power of the natural world, there remains nonetheless a persistent voice of concession to the authoritative view of the otherworldly control of divine omnipotence: the position of natural self-sufficiency is never permitted fully and unequivocally to assert itself. The period's most powerful expression of self-sufficiency, as we saw in Chapter 4, can be found in the desperate words of Satan in *Paradise Lost*, who denies the very act of his creation:

> remember'st thou
> Thy making, while the Maker gave thee being?
> We know no time when we were not as now;
> Know none before us, self-begot, self-rais'd
> By our own quick'ning power.
>
> (5.857–61)

Satan, one of the poem's prime expositors of monistic vitalism, exploits the logical inconsistency that plagues the coexistence of angelic self-movement and God's alleged omnipotence.[55] Milton, for his part, is usually content to suggest that God's original endowment of matter with the power of movement had merely released the Deity from any immediate and ongoing responsibility for his creation, a position that would have been congenial to the implicit theology of Cavendish's science. But there are moments in Cavendish's work in which the logic of her position seems to push her even further than the official compromise at which Milton had arrived, bringing her dangerously close to the radical atheism of Milton's Satan: "every part and particle of Nature has the principle of motion within it self, as consisting all of a composition of animate or self-moving Matter; and if this be so, what need we to trouble our selves about a first Mover?"[56] Although the imme-

---

55. Kendrick discusses Satan's monism in *Milton*, p. 175.

56. Cavendish, *Observations upon the Opinions of Some Ancient Philosophers*, p. 3, in *Observations upon Experimental Philosophy*. As early as her *Poems, and Fancies*, Cavendish imagined the sufficiency of matter not only to constitute but even to create the known world. Her atheistic cosmogony is articulated best in the volume's first poem: "Small *Atomes* of themselves a *World* may make," p. 5 of *Poems, and Fancies*.

diate object of Cavendish's doubt in this passage is not the Christian God
but Aristotle's Prime Mover, the argument is readily transferable from a
classical to a Christian context. We can hear in just such moments in Cavendish's prose Satan's liberatory recognition that a consciousness of self-
animation can release one from troubling about createdness.

As we saw in Chapter 4, the concerns informing Satan's denial that the
so-called Maker gave him "being" are primarily political: Satan is adducing
evidence to justify his resentment of the absolute tyranny he envisions as
the heavenly polis. As we attempt to discern the function of Cavendish's
claim of self-moving matter, we may want to keep in mind the political
nature of the narrative that supports Satan's boast of self-generation. Cavendish's philosophy differs from nearly all contemporary scientific accounts
of natural change in the nature of the organizational structure that underlies
it. Like the primeval Satanic heaven with its self-creating angels, the Cavendishean universe is marked by the shocking absence of any single regu-
latory power or absolute center of command. Catherine Gallagher has argued
for the centrality of the figure of absolute sovereignty for Cavendish's liberatory construction of female subjectivity.[57] We must not, however, derive
from this important insight a notion of Cavendish's commitment to an organizational principle of centralized absolutism; it is precisely the attribution
of sovereignty to each individual element in a society that marks the dialectical construction of liberalism at its mid-seventeenth-century inception. Margaret Cavendish's vitalist utopia, more powerfully than the official vitalisms
of Harvey or Milton, is structured as an unabashedly liberal system, a system
of disseminated sovereignty devised quite specifically to counter the authoritarian organization of the leading theories of her day.

It is the antiauthoritarian, republican structure of her vitalist philosophy
that seems to have shocked her contemporary Cudworth: "And to say, that
these innumerable *Particles* of *Matter*, Do all *Confederate* together; that is,
to make every Man and Animal, to be a *Multitude* or *Common-wealth* of
Percipients and Persons as it were clubbing together; is a thing . . . *Absurd*
and Ridiculous."[58] Rather than the absolute subjection of a multitude of inert
particles to a univocal higher power, Cavendish's universe is for Cudworth
a *"Common-wealth"* of perceptive, thinking beings, a liberal form of organization that the Royalist Cudworth would no doubt find impracticable and
absurd in whatever organizational sphere it was imagined to apply. In his
stinging evaluation of the confederacy of material percipients in what Cav-

57. Catherine Gallagher, "Embracing the Absolute: The Politics of the Female Subject in
Seventeenth-Century England," *Genders* 1 (1988): 24–39.

58. Cudworth, *True Intellectual System*, p. 839.

endish calls the "Common-wealth of the body,"[59] Cudworth is no doubt
alluding to the political failure that was the republican Commonwealth. But
Cavendish's commonwealth of clubbable rational particles is founded on an
organizational principle never actually instituted in the political sphere during the years of the English republic. Although her thesis of the perceptive
self-motion of rational matter has few parallels in the broad spectrum of
seventeenth-century natural philosophical speculation, we nonetheless recognize the organizational logic that subtends her natural philosophy. Her
decentralized yet hierarchical system of material government resembles
nothing so much as the republican Puritan ideal of the rule of the godly
few.[60] In *The Readie and Easie Way to Establish a True Commonwealth*,
written just weeks before the monarchic Restoration in 1660, Milton tried
to halt the drift toward a restored monarchy, "which the inconsiderate multitude are now so madd upon" (6:134). He proposed the establishment, as
we observed in Chapter 4, of a self-governing community of regenerate saints
whose authority derived solely from their self-sufficient powers of reason.
The true commonwealth for Milton could exist only under the stewardship
of a "full and free Councel . . . where no single person, but reason only
swaies" (6:121–2).

Infusing an entire class of material parts with rational mind, Cavendish
aligns herself, if not with the most radical dissenters of the Vitalist Moment, then with those republicans, like the Milton of the late 1650s, who
attempt to balance their antimonarchic politics with their distrust of the
masses. Her claim for the knowledge and infallible reason with which rational matter is endowed displaces onto the sphere of natural philosophy
that radical Protestant doctrine that had been so effective in the 1640s in
arousing opposition to the English prelacy: the doctrine of the priesthood
of the believer, or what Merritt Hughes called "an aristocracy of grace."
The attraction of such a rational oligarchy for both Cavendish and Milton
is clear. The society as a whole may be organized hierarchically, but the
perpetual senate of the rational elements of society need recognize no authority outside themselves, since the very premise of an absolute rationality
renders the existence of a single authoritative ruler unnecessary. The only
tenets to which Cavendish's rational particles are obliged would seem to
be those two principles Milton honored in *The Readie and Easie Way*:
"full liberty of conscience and the abjuration of monarchy." With the ca-

59. Cavendish, *Philosophical Letters*, p. 366.
60. On the compatibility of hylozoism and republicanism in the political philosophy of James
Harrington, see William Craig Diamond, "Natural Philosophy and Harrington's Political
Thought," *Journal of the History of Philosophy* 16 (1978): 387–98.

pacity to govern themselves, these infinitely wise particles are the architects of a true commonwealth.

Although, as I have noted, Cavendish briefly entertains a thesis of the essential equality of all material parts,[61] she is most often committed to structuring her animate universe around an aristocratic hierarchy of rational and sensitive matter. I want now to explore in some depth the nature of her inveterate hierarchism, because I think it is precisely the hierarchical aspect of her decentralized organizational structure that permits her natural philosophy to function as a discourse we can identify as feminist. Hobbes in his political philosophy had pointedly rejected Aristotle's naturalization of aristocratic rule; he scoffed at Aristotle's enslavement of "those that had strong bodies, but were not Philosophers as he."[62] Hobbes proposed in opposition to Aristotelian aristocracy a radical and curiously argued premise of equality.[63] But in Cavendish, as in Milton, we find a stubborn allegiance to Aristotle's justification of nonmonarchical social hierarchy, one that is founded specifically on the subjection of what Aristotle calls "bodily power" to "intelligence." For Aristotle, "the element [of society] which is able, by virtue of its intelligence, to exercise forethought is naturally a ruling and master element; the element which is able, by virtue of its bodily power, to do what the other element plans, is a ruled element, which is naturally in a state of slavery."[64] By virtue of its intelligence, Cavendish's rational matter designs and plans the movements of the laboring sensitive matter, just as Milton's saints must guide and direct the unregenerate, those monarch-loving English brutes he called the "misguided and abus'd multitude" (6:149). It is in opposition to the Hobbesian precept that the successful exercise of force always determines the rule of right that both Milton and Cavendish fashion a neo-Aristotelian organization that places the mantle of power on a rational elite, on a godly few whose authority derives not from strength but from a conscientious desire solely for liberty and self-government.

In aligning Cavendishean science with a certain strain of idealist Puritan politics, I do not at all want to suggest that Cavendish's philosophical vitalism betrays anything like a secret or coded sympathy with the republican politics of the Interregnum. There is no indication that the duchess ever questioned the ideology of aristocratic privilege or that she would ever have compro-

61. See note 38, above.

62. Hobbes, Leviathan, p. 211.

63. Hobbes offers as proof for his contention that "Nature hath made men . . . equall" the fact that "the weakest has strength enough to kill the strongest, either by secret machination, or by confederacy with others" (Leviathan, p. 183).

64. The Politics of Aristotle, ed. and trans. Ernest Barker (Oxford: Clarendon, 1946), p. 3 (1252a).

mised her commitment to the stability of the Stuart monarchy.[65] She seems to have made a concerted effort to distance herself from the political and theological forces that motivated the monistic vitalism of a radical such as Overton or Winstanley. In her *Observations upon Experimental Philosophy*, she condemns, without the slightest hint of irony, those

> Natural Philosophers, who by their extracted, or rather distracted arguments, confound both Divinity and Natural Philosophy, Sense and Reason, Nature and Art, so much as in time we shall have rather a Chaos, then a well-order'd Universe by their doctrine. . . . such Writers are like those unconscionable men in Civil Wars, which endeavour to pull down the hereditary Mansions of Noblemen and Gentlemen, to build a Cottage of their own; for so do they pull down the learning of Ancient Authors, to render themselves famous in composing Books of their own.[66]

The "extraction" of the political or theological implications of natural philosophy can, as Cavendish knew very well, constitute a powerful act of rebellion. The explicit extension of monistic vitalism to the vital interactions of people in society could indeed be tantamount, as we know from the work of Winstanley, to the action of "those unconscionable men in Civil Wars, which endeavour to pull down the hereditary Mansions of Noble-men." The subversive exploitation of the organizational analogy could actually fuel a politics that sought to destroy the cherished network of political privileges founded on heredity, property, and tradition.

The Duchess of Newcastle, I believe it is safe to claim, did not, like Winstanley or even Milton, seek to "pull down the hereditary mansions of Noblemen." But I think Cavendish's consistent use, in this very text, of the architectural justification of her idiosyncratic science of motion suggests that she was herself deeply invested (*pace* Virginia Woolf) in building a philosophical "Cottage" of her own. We have the evidence of her own *Observations upon the Opinions of Some Ancient Philosophers*, appended to this same volume, to suggest that she is as guilty of self-assertion as those most radical "natural philosophers" who "pull down the learning of Ancient Authors, to render themselves famous in composing Books of their own." Cavendish may struggle to dissociate herself from a republican politics, but she evinces throughout her work a powerful attraction to a republican organizational logic and to the unbridled self-assertion her age came to associate with republican ambition.

65. Schiebinger, *Mind Has No Sex?* p. 58. See Cavendish's expression of monarchism in the story "The She-Anchoret," in *Natures Pictures*, pp. 634–35.
66. Cavendish, *Observations upon Experimental Philosophy*, sig. c2[r].

We are left now to speculate why the fiercely Royalist Duchess of New-castle, not unlike the fiercely Royalist Dr. Harvey fifteen years before, would appropriate for her natural philosophy the organizational lineaments of a political agenda to which she could not have been more opposed. Cavendish, I would argue, deploys the Nonconformist rhetoric of rational choice neither for its political association nor for its attachment to a set of individualist religious principles. She invokes the structure of the Dissenters' rational uto-pia for its rhetorical power to suggest a radical program of emancipation. It is a revolutionary zeal she is conjuring, the intellectual daring (though not the ideological content) of the Parliamentary opposition to that institution of power referred to so often by radical Puritans as "tyranny." It is not of course the tyranny of a restored Stuart monarch that Cavendish opposes. It is the far less localized tyranny of a system of sexual subjection, a cultural institu-tion that had found an authoritative scientific justification in the new mech-anist belief in the rule of brute force. The egalitarian infusion of matter with motion, or the conjecture of innumerable agents of infallible reason, may have emerged originally from a radical politics of dissent. But this new liberal logic could be made to speak to a potentially unlimited range of ideological concerns. Later in the century, the writer Mary Astell complained that Mil-ton was never able to extend his critique of tyranny from the political to the domestic sphere: "Patience and Submission are the only comforts that are left to a poor People, who groan under Tyranny, unless they are Strong enough to break the Yoke. . . . Not *Milton* himself would cry up liberty to poor *Female Slaves*, or plead for the Lawfulness of Resisting a Private Tyr-anny."[67] Astell makes explicit the specifically masculinist assumptions behind the Miltonic ideal of freedom. But her comment also points to the potentially fruitful discursive alliance between the liberatory goals of a Puritan revolu-tionary such as Milton and the plea for female emancipation by a Royalist intellectual such as Margaret Cavendish.

Despite the absence in Cavendish's theory of any explicit organization of material hierarchy into male and female components, we can nonetheless detect in her rhetorical figuration of the hierarchy of animate matter an important reinscription of the category of gender. The role of brute force or physical strength is assumed in her science by the sensitive particles of matter. Given Cavendish's inveterate identification of physical strength with maleness, it is difficult not to imagine that these lumbering, laboring, obedient, material particles are to be seen, if only for their muscle, as masculine. Cavendish had complained in 1655, we remember, that men

---

67. Mary Astell, *A Serious Proposal to the Ladies for the Advancement of their True and Greatest Interest* (London, 1696).

alone are "able to endure Hard Labour" and "to bear Weighty Burthens." But as her revitalized natural philosophy suggests, she had by 1663 transformed what was previously considered a physiological advantage into a distinct liability. Just as those possessed of bodily "power" in Aristotle's *Politics* were subjected to the state of slavery, so Cavendish's sensitive particles of matter are compelled to serve as "Work-men, being always busily imployed, as in Removing, Lifting, Carrying, Driving, Drawing, Placing, Digging, Cutting, Carving, Forming, Fixing, Measuring, and millions the like."[68] Their strength, however necessary, is not their own and is eternally at the command of rational matter. For its part, the rational element of matter is guaranteed superiority by virtue of its intelligence and, additionally, its *lack* of bodily strength; rational matter "is so pure and free, as it cannot be so painfull a Labourer as to work on the gross Unanimate matter."[69] Retiring to its cozier realm of purity and freedom, rational matter bears an indisputable rhetorical affinity with Cavendish's figuration of women, whom Nature has fashioned of a purer and more refined substance than the crude matter of men, in order to make "them neerest to resemble, Angells, which are the perfectest of all her Works."[70] In the world of late Cavendishean physics, the lack of workmanlike masculine strength is a sign not of inferiority but of aristocratic privilege, of a purity and freedom that guarantees for rational matter its authority. Cavendish's science is, to be sure, a sincere exercise in organizational physics; but this physical vision functions most powerfully as a utopia of female rule.

We may begin to understand the process by which Cavendish has gendered the relationship between rational and sensitive matter by examining an early adumbration of this problem in her essay "Noble Souls, and Strong Bodies" in *The Worlds Olio*. Brooding as so often on the relative merits of the male and female sexes, Cavendish declares, "if I were to choose a Sex, I had rather be a Pigmy, stuft with rational spirits, than a Giant empty thereof."[71] This representation of a choice of gender seems itself to enact the curious alchemy by which Cavendish's interest in female emancipation becomes transmogrified into a natural philosophy of rational matter. We expect, of course, in reading the initial condition of this remarkable sentence that Cavendish will declare that she "had rather be" one sex or the other. But the anticipated distinction between male and female slips without warning into a distinction between a rational and a "sensitive" (or physical) crea-

---

68. Cavendish, *Philosophical and Physical Opinions*, p. 19.
69. Ibid., p. 3.
70. Cavendish, *Worlds Olio*, sig. A5ᵛ.
71. Ibid., p. 216.

ture. The awkwardness of the sentence's lapse in logic seems almost to announce the urgency with which Cavendish desires to replace the category of gender with the category of rational capacity. But though she chooses to replace the standard of sex with the standard of reason, the new distinction between a "Pigmy, stuft with rational spirits," and a "Giant empty thereof" works to reinscribe, in a new key, the antinomies of sexual difference. The distinction between pygmy and giant introduces the categories of bodily size and strength, the very categories that are constitutive for Cavendish of sexual difference itself. Size and strength have within the utopian logic of Cavendish's natural philosophy an inverse relation to reason. It is clear that to be the rational pygmy in this world, as to be an almost bodiless rational part of matter in the universe of microparticles, is to enjoy the rights and privileges of a newly dominant, and decidedly female, governing class.

Every movement in nature is for Cavendish the product of a complex architectural project. Her architect atoms are forever engaged in the process of building a secularized and feminized version of the radical Puritans' holy community, an ideal commonwealth that was, at least for Milton, grounded on the subordination of physical force to the force of reason. Although it is doubtful that Cavendish ever read Milton's epic, we can hear in her chemistry a utopia akin to the passive revolutionary stance Adam anticipates at the end of *Paradise Lost*: "by small / Accomplishing great things, by things deem'd weak / Subverting worldly strong" (12.566–68).[72] Adam looks ahead here to the verse in First Corinthians that Milton quotes in *A Treatise of Civil Power* (6:22): "God hath chosen the weak things of the world to confound the things which are mighty" (1 Cor. 1:27).[73] It is this promise of the ultimate triumph of the weak that has attracted Cavendish to the Puritan commonwealth of reason; she appropriates for her utopian science its rejection of outward force and its subordination to contemplative reason of the traditional authorities of civil law and civil power. In organizing her scientific vision as a radical politics, Cavendish establishes the metaphysical foundations of a universe in which women might not be kept like birds in cages or compelled by subjection into an irrational idiocy. As the very embodiments of reason itself, they will instead govern everything around them by governing themselves.

72. In his late pamphlet *A Treatise of Civil Power*, Milton meditates on the role of military strength in his rational theocracy, concluding that God "hath not chosen the force of this world to subdue conscience and conscientious men, who in this world are counted weakest." God has chosen "rather conscience, as being weakest, to subdue and regulate force, his adversarie, not his aide or instrument in governing the church" (6:22).

73. Milton had invoked this verse from Paul as a consolation for his blindness, in 1651, 1654, and 1656, according to Parker, *Milton*, 1:389.

Their Destiny Their Choice

Cavendish had in Hobbes's *Leviathan* the preeminent model of a theory of human society that presented itself as a logical extension of a theory of natural causation. We have examined how her natural philosophy represents a new application of this Hobbesian procedure, an attempt to reconceptualize the organizational underpinnings of the social phenomenon of sexual subjection. It is important to note, however, that Cavendish is not, like Hobbes, explicit about the analogical connection between the worlds of material and social organization. However ideologically resonant her philosophy may be, her account of matter in motion is never openly made available for a corresponding prescription for social interaction; her discourse simply never descends from the airy realm of scientific theory to the mundane world of political philosophy.[74] We might be tempted to ascribe this absence of overt politicization to the impossibility during this period of giving public voice to sentiments so subversive of patriarchal rule. But Cavendish's extrascientific essays are remarkable, I should think, precisely for the fearless volubility with which they engage the topic of the subjection of women. Given the astonishing forthrightness that characterizes nearly all her direct considerations of women in society, it is impossible not to find somewhat puzzling the absence here of the Hobbesian gesture of pressing a natural philosophy into service as the theoretical basis for a philosophy of the polis. I want now to investigate the failure of Cavendish's texts to extend their revolutionary conclusions from the world of material particles to the world of human beings. This failure, I propose, may bespeak a large and disconcerting shadow of defeatism that darkens Cavendish's otherwise illuminated discourse of rational feminine choice.

We have up to this point explored the liberatory feminist implications of only one aspect of Cavendish's explanation of mundane physical phenomena: the internalist perspective on bodily motion whereby movement is the consequence of the choices enacted by the members of the material organization within a given material body—in a hand, say, or a billiard ball. From the privileged theoretical perspective into the invisible world within every material object, the implicitly feminine rational part of the particulate society enjoys an undisputed power and pride of place. But Cavendish, as we have seen, in maintaining her focus on the microstructural dynamics of internal volition, must deny any immediate causal efficacy to the bodies and forces—

74. Margaret J. M. Ezell, *The Patriarch's Wife: Literary Evidence and the History of the Family* (Chapel Hill: University of North Carolina Press, 1987), p. 113, criticizes Cavendish for her refusal to "offer any formal plan to correct this fault" of the sexual injustice she exposes.

plain enough to the naked eye—residing outside a material object. It is the struggle, I believe, with the unavoidable externalist attention to bodily collision that exposes the real, perhaps insurmountable, obstacles confronting Cavendish's feminist physics. I want now to draw our attention to a rare instance in her work in which the subject of forces external to a material body is explicitly engaged. In an essay appended to her *Observations upon Experimental Philosophy*, Cavendish attempts to reinterpret the paradigmatic event of the collision of two bodies as a nonmechanical sequence of actions: " 'Tis true, one part may occasion another by its outward impulse or force, to move thus or thus; but no part can move by an others motion, but its own which is an internal, and innate motion."[75] One body, she argues, may "occasion," but never cause, the motion of another body. The distinction Cavendish puts forward here between occasion and cause is crucial for the success of her science, and I propose that we scrutinize the sequence of action between the forceful "occasioning" of a body's motion and its subsequent, freely willed "self-moving."

Cavendish is clearly concerned to grant priority to a body's own "internal and innate motion" over the forceful pressure of bodies external to it. But the *temporal* priority in this chain of causation necessarily belongs to the impulsive act of occasioning. And the question Cavendish's philosophy never sufficiently answers involves the nature of the causal relation between the violent crash of occasion and the internal and innate motion that seems almost invariably to follow it. An animate body's motion can seem as much an act of resignation to the pressure of external forces as it seems a positive and freely willed self-action. The Cavendishean choice that motivates motion in fact resembles the heavily circumscribed form of choice captured in the famous oxymoron of Marvell's *Upon Appleton House*. Lord and Lady Fairfax, in conceding to the public pressures that necessitate the marriage of their virgin daughter, must make "their *Destiny* their *Choice*" (744). In doing so, however, they must first accede to an entirely new definition of *choice*, a *choice* whose semantic value has been narrowed from a positive selection among genuine alternatives to the voluntary resignation to a higher power.

Throughout her later works, Cavendish deploys the rhetoric of free will to combat the oppressive power politics of mechanistic determinism. But this language of free and rational choice threatens continually to slide, as it did for the Fairfaxes, into a related, but ultimately very different, language of political obligation. The ethical discourse of *choice* functions in many ways as Cavendish's sanguine reconfiguration of the political discourse of *consent*.

---

75. Cavendish, *Observations upon the Opinions of Some Ancient Philosophers*, p. 3, in *Observations upon Experimental Philosophy*.

The rational parts of matter belonging to the weaker of two colliding bodies are not so much in a position to choose their course of motion as to consent to, or even comply with, the motion established by the stronger body. The distinction between choice and consent cannot be overlooked, and it may be useful in this context to avail ourselves of a twentieth-century analysis of the role of consent theory in seventeenth-century political philosophy. Although, according to Don Herzog, the discourse of consent emerged rhetorically in the theoretical consideration of the relations of putatively equal members of society, the liberal rhetoric of consent invariably masked the facts of inequality structuring that society:

> Consent doesn't mean that relations of hierarchy and deference were wholly voluntary, that some contractual act among equals somehow preceded and legitimated daily social relations. Hierarchy and deference were already firmly in place, and individuals had nothing resembling a choice about them. Their consent though shows that they accede to the proposed reading of what the hierarchy amounts to, that the authorities have identified the rightful leader.[76]

Hobbes's genius lay in part in his ability to refigure every act of obligation, or even of coercion, into a drama of tacit consent, "there being no obligation on any man, which ariseth not from some act of his own."[77] But from any meaningful practical standpoint, as Herzog suggests, consent can be less one's own choice than one's concession to the authoritative choice of someone else. Consent is always in some way occasioned (perhaps even caused) by a prior condition of obligation, since one is in actuality called on to grant consent only if one is already in a position of relative powerlessness or in a posture of deference on the hierarchical ladder.

As we might expect, Cavendish works hard to scuttle the possibility that the free will of her theory is in fact forced consent, convinced, at least in a discussion of the hypothesis of atoms, that "there must necessarily be as much Liberty and Power in every atome to Disagree as to Agree."[78] But the depressing regularity with which parts of matter "agree" to move when struck by the parts of matter of other, stronger bodies suggests all too forcefully that there is probably *not* as much liberty and power in animate matter to disagree as to agree. The free will Cavendish heroically attributes to ma-

---

76. Don Herzog, *Happy Slaves: A Critique of Consent Theory* (Chicago: University of Chicago Press, 1989), p. 198. Herzog provides a useful account of "two concepts of consent" on pp. 196–200.

77. Hobbes, *Leviathan*, p. 238.

78. Cavendish, *Philosophical and Physical Opinions*, sigs. c2^r–c2^v.

terial bodies quietly slides, on reflection, into the more deferential realm of assent and acquiescence.

The degree of liberty and power we can attribute to bodies of matter depends almost entirely on whether we view them, as the scientific members of the Royal Society might have viewed them, from the outside, as objects in the empirical world; or whether we view them, as Cavendish chose to, from the inside, as free inhabitants of the invisible world of material parts. The binary structure of these competing perspectives no doubt strikes us as familiar: the tension between the empirical perspective on outward objects and the rationalist perspective on the inner world of material parts reproduces almost exactly the conceptual antinomy of the public and the private spheres of human activity. It is, I would argue, the period's culturally sanctioned opposition between public and private realms that provides the best analogue for the contradictions fissuring Cavendish's science of rational agency.

We observed in the chapters on Marvell the degree to which the notion of the public and the private spheres had acquired in the seventeenth century a strongly gendered set of associations. An increasingly impassable divide seems to have segregated the masculine polity from the feminine household. And this intensified opposition made possible the conceptual establishment of a realm, the private sphere, in which women were seen actually to govern. The virginal Mary Fairfax, for example, was figured as commanding a formidable sway over the elements of landscape within the private boundaries of her family's property. Women could be seen to enjoy pride of place in the government of the household, much as Cavendish's rational atoms govern that inner world of the parts of matter. But the rhetoric of sexual segregation, as often as it suggested the possibility of parity, invited the diminution of the private realm of feminine rule when viewed aside the public realm of traditionally masculine dominance. Mary Fairfax may command the awe of the "lesser *World*" within the confines of the family estate, but we cannot but assume that she is impotent to change the greater world of a nation divided by war. An acknowledgment of the limitations besetting the separate sphere of female sovereignty is nowhere voiced so clearly as in *The Blazing World*, the utopian romance Cavendish appends to her *Observations upon Experimental Philosophy*. There the Empress has command not only over a private world within her imagination, but also over an actual polity in the form of the planet of the Blazing World. The imaginative world, interestingly "composed only of the rational" parts of self-moving matter, she is able to dissolve and reconstitute at will (pp. 188–89). The actual public world under her control, however, proves much less tractable: "she could hardly" make "some alterations in the Blazing World she lived in" (p. 189). And although the narrator informs us that the Empress's public powers were constricted for the happy reason that her Blazing World "was so well ordered that it could not be mended," a civil war nonetheless breaks out, and the few alterations the Empress is able to make must, by the work's conclusion, be repealed (pp. 189, 201). The feminine parts of rational matter that demand such respect in the inner world of Cavendishean science are enmeshed in a similar tangle of limitation and power. However enviable their authority within the domestic enclave of the microworld of material parts, they seem more or less powerless to alter or avoid the brute laws that continue to govern the collision of outward objects.

I suggested at the beginning of this chapter that Margaret Cavendish can be seen to raise to the level of formal articulation the undisclosed dream of a liberatory natural history that, I speculated, was bestowed on Eve in the last two books of *Paradise Lost*. I would like to suggest now that we listen to Cavendish not only for the voice of Eve's liberatory vitalism but also for the more compromised voice of the Eve constrained by a patriarchal paradise. On the day of her creation in *Paradise Lost*, basking in the self-sufficiency afforded by her sympathetic reflection, Eve is commanded by an invisible spirit to take her place as a subordinate in the more public world of human interaction. She obeys that voice and finds in Adam her "Guide / And Head." In a passage remarkable for its demystification of the poem's official ideology of free will, Eve recalls later the moment at which she "chose" to follow Adam as wife: "thy gentle hand / Seiz'd mine, I yielded" (4.488–89). Drawing our attention to that mysterious moment between Eve's physical contact with Adam's more powerful hand and her positive decision to yield to that power, Milton permits us to see the troubling dialectic of coercion and consent, however gentle, that seems necessarily to underlie the principle of freedom in a hierarchical society. Eve's is that peculiarly stoic form of freedom any upper-class seventeenth-century woman might have found herself exercising as she consented to give away her hand in what was essentially a prearranged marriage. In the rhythm of the seizing and yielding that constitutes the first act of human contact, we recognize the "occasioning" and "self-acting" that make up the essence of Cavendish's theory of motion. Reflecting, perhaps, in all its contradiction the most extravagant form of liberty available to a woman in the seventeenth century, the mode of choice open to Cavendish's rational matter, as to Milton's Eve, brings us disconcertingly close to Hobbes's determinist position that "*Liberty* and *Necessity* are Consistent."[79] There is no question that Cavendish and Milton are both under pressure to avoid Hobbes's constrictive conclusion; they

79. Hobbes, *Leviathan*, p. 263.

struggle to carve out for their rational agents a space of freedom that allows as much liberty to disagree as to agree. But we have in the words of Eve herself, as she remembers her obedience to the power who first brought her to Adam, the proper translation of this "choice" into a wearier state of helpless resignation: "What could I do, / But follow straight, invisibly thus led?" (4.475–76).

We must of course concede that Cavendish's hylozoic philosophy of matter performs a radical inversion of the ideological values traditionally assigned to matter and to force, an inversion that points to a utopian recharting of the relation of women to men. But we should be instructed by the eerie proximity of her own logic of agency to the type of female agency articulated by the Eve of John Milton. There is an important way in which Cavendish's revolutionary theory of animate matter simply replicates in the utopian discourse of material relations those structures the duchess had found so oppressive in the experiential world of social relations. Nature had traditionally been seen as a feminine component of a larger, and implicitly masculinist, universe, an all-encompassing conceptual system in which Mother Nature played a subordinate role either to a Heavenly Father or to any of a number of disembodied forces of motion and strength. And although Cavendish is more or less successful in bracketing the question of a bullying, supervisory God, her science still falls prey to a theory of causation, like Hobbes's, whose focus is trained on the physical impact of contiguous bodies.

Much of Cavendish's scientific writing, I would argue, is devoted to thwarting the Hobbesian conjecture of the ultimate priority of force. And her strategy in this project has been to effect a radical shift in focus: Cavendish has transposed the analysis of feminine nature from the implicitly "public" context of the external impact of outward objects to the "private" context of the household organization within objects. But it should now be possible, I think, to understand the failure of her natural philosophy to assert itself as a liberatory social philosophy. The conflicting ideological implications of Cavendish's science (the confusion produced by the clash of internalist and externalist physics) reproduce in all their complexity the conflicting pronouncements on women's rights and abilities that strain the coherence of Cavendish's cultural commentary. Like the explicit feminism of her essays and epistles, the implicit feminism of her science calls for a sweeping liberation from the constraints of patriarchy at the same time it works to confine the exercise of female power to the home. The ideologically loaded rhetoric of her scientific theory seems ultimately to reinscribe, however inadvertently, the dynamics of resignation and assent that it may very well have been fashioned to defeat.

Edmund Waller, it is believed, is responsible for writing these lines about

Cavendish's philosophy: "New Castles in the air this Lady builds, / While nonsense with Philosophy she guilds."[80] This play on the duchess's name was one that followed her, with innumerable variations, throughout her career.[81] Cavendish herself engages the wordplay in a letter, from *CCXI Sociable Letters*, that she writes to a fictional friend. Having been accused, she writes, of "Imploying my time onely in Building Castles in the Air," she responds not with a denial but with a defense of her construction of such castles: "as for the Minds Architecture, as Castles in the Air, or Airy Castles, which are Poetical Conceptions . . . they will be more lasting than Castles of Wood, Brick, or Stone."[82] As we have observed, the dominant conceptual paradigm for the Cavendishean theory of motion is architectural: the rational matter, when viewed with an eye to its matriarchal control over the other types of matter, is continually designing the elaborate structure of every physical movement. The edifice of Cavendish's science of motion seems to provide the foundation for an entirely new society, a society in which every woman's home might be her castle. But there is a marked resistance in Cavendish's work to imagining the life of the rational parts of matter outside the castles of their own design. The structural outlines of her science are confined to the interior of outward objects, much as her own philosophy remains nestled in the secluded organizational domain of scientific, and not social, theory. Margaret Cavendish is scrupulously careful to keep her castles in the air from establishing their status as anything but the "Minds Architecture."

Having examined the disquieting tones of resignation that seem to color her theory of voluntary rational action, we can now see in its proper light Cavendish's rhetorical alignment of her revolutionary science with the revolutionary utopias fashioned by Puritan radicals such as Milton. She conjures the dissenting fury of some of her age's most daring political voices, but her appropriation of the rational utopia of a postrepublican Puritan such as Milton is also fraught with a less optimistic set of associations. The Puritan fight for a holy community was little, by the time of the Restoration, but a deeply compromised position of dissident political sentiment.[83] And the ghost of the

---

80. This couplet was inscribed on the flyleaf of Edmund Waller's copy of *Philosophical and Physical Opinions* (1663), now in the Huntington Library (shelf mark 120156). See Perry, *First Duchess of Newcastle*, p. 179n.

81. I am indebted to James Fitzmaurice for this point, and for the reference to *Sociable Letters*.

82. *CCXI Sociable Letters* (London, 1664), pp. 226–27.

83. Richard L. Greaves, in *Deliver Us from Evil: The Radical Underground in Britain, 1660–1663* (New York: Oxford University Press, 1986), documents the resiliency of radical political discourse in the early years of the Restoration. Although a revolutionary sensibility may have continued well into the 1660s to pose a threat to the new Stuart regime, it is largely, I believe,

failure of this idealist Puritan rhetoric haunts the duchess's otherwise exuberant picture of the newly inverted hierarchy structuring her commonwealth of matter. Cudworth, writing after the Restoration, could refer to Cavendishean philosophy as a "Ruinous and Desperate Cause"; he implicitly allies the revolutionary matter of her dissenting science with the ruin and desperation to which the supporters of the Good Old Cause had been driven.[84] Cavendish's invocation of the defeated cause of the vitalist revolutionaries signals perhaps a related sense that a utopia of female volition—like the true commonwealth Milton foresaw, like the communist state of the Diggers, like the green age of Marvell's idle pastorals—might never materialize.

---

the official defeat of the Good Old Cause that informs Cavendish's own engagement of a political radicalism inescapably tied to an earlier epoch.

84. Cudworth, *True Intellectual System*, p. 145.

Adamant Liberals:
The Failure of the Matter of
Revolution

There was, however, another revolution which never happened.
— CHRISTOPHER HILL, *The World Turned Upside Down*

It is possible there was never a point at which any of the vitalists we have considered was secure in the faith that a natural philosophy of self-moving matter could prompt a reorganization of society or polity. We have not seen represented a consistent or uncompromised subversion of authoritative discourse by the vitalist rhetoric of agency and organization. We have seen instead how each of these writers has inscribed within his or her work a sense not only of vitalism's attraction but, too, of its inevitable defeat. The natural philosophy of vitalism did not, we know, prove sufficiently strong to withstand its dissipation by the mechanistic philosophies of matter that came to dominate the Scientific Revolution. It may not have been strong enough to survive the doubts of the vitalists themselves. But liberalism, that organizational abstraction I have provisionally associated throughout this book with vitalism, clearly did move on to establish itself as an authoritative set of discursive positions. I want now to consider why vitalism, as a natural philosophy, failed to sustain on ontological grounds the general faith in independent agency and decentralized organization that would come to be known as liberalism. With an eye to the relation of the political and rhetorical energies with which this book has been concerned, I submit an explanation for vitalism's inability to deliver to modernity the liberal individual. The failure of the vitalist revolution can be seen both as a reflection of and as a response to an aggregate of failures occurring disconcertingly early in the English Revolution.

Harvey, Winstanley, Marvell, Milton, and Cavendish have all encoded in their work, with differing degrees of self-consciousness, representations of matter in motion that present themselves as elegies for the vitalist ideal. These expressions of vitalism's failure, I believe, surface primarily in the shape of two distinct but closely related elegiac forms: the figurations of vitalism's death as sacrificial and as degenerative. We can begin our analysis of this movement's abortive beginnings with a look at two texts from the Vitalist Moment that neatly exemplify the first, sacrificial, mode in which the untimely demise of this promising natural philosophy is represented: Marvell's "Nymph Complaining for the Death of Her Faun," surely the finest and most sustained elegy for vitalism, and one of the works with which this book began, Harvey's *De circulatione sanguinis* of 1649. Marvell and Harvey present in usefully emblematic form the conflicted dynamics of vitalist expression that characterizes all the texts we have considered. In his 1651 account of the examination of the soon-to-be Viscount Montgomery, Harvey, as we saw in Chapter 1, would represent as a miraculous discovery his disestablishment of the centralized organization of the body. As if drawing on Paul's vitalist image of the parturient earth, "groaning in labor pain," Harvey would cast his account in that literary mode best suited to such miraculous deliveries, the mode of romance. But we have in the *Circulatio* a very different emplotment of a closely related physiological discovery. Here Harvey figures as tragedy—if only for a moment—the discovery of vitalist agency that he represents throughout the rest of his work as a romance of truth. This brief narrative functions for Harvey as a tragic complement to his romance account of Montgomery's insensitive heart; it can function as well, as we will see, as a fortuitous companion of Marvell's "Nymph Complaining."

In 1649, Harvey does not open his treatise, as in 1628, with a dedication to the king. But Charles is nonetheless present in this later, vitalist reading of the blood's circulation, having borne witness to another of Harvey's ideologically overburdened experiments. In the *Second Essay to Riolan*, which comprises half of the *Circulation of the Blood*, Harvey documents a second demonstration performed before the king, this one proving by a very different means the fact that the mainspring of the circulation lay not in the heart but in the blood:

> In the internal jugular vein of a live Doe, which I laid before a great part of the Nobility, and the King my Royal Master standing by, which was cut and broke off in the middle: From the lower part rising from the Clavicule, scarce a few drops did issue, whilst in the mean time, the blood with great force, and breaking out of a round stream, ran out most plentifully downwards from the head through the other orifice of the vein.[1]

1. Harvey, *Anatomical Exercises*, ed. Keynes, p. 170.

The dissection of the doe's jugular vein is performed to display the autonomous movement of blood toward the heart: the desultory dripping of blood leaving the heart cannot compare in strength to the "great force" of blood moving up from the extremities. The agent behind this motion is identified not as the heart's forcible impulsion of arterial blood, or its magical attraction of blood from the veins, but the "continuall and great flux" of the blood itself. Harvey recounts his discovery with the observational precision for which he is justly celebrated. But surely it is impossible not to read into this revelation of vitalist agency the king's proleptic experience of his own beheading, an event at which perhaps, too, "the blood with great force . . . ran out most plentifully downwards from the head." The Royal Physician has exposed the neck of one of the king's deer in the presence of the king himself, whose own jugular vein would be severed in the very year this treatise is published.

When Marvell's Nymph complains that the "wanton Troopers" have shot her fawn, the intrusion of midcentury English soldiers on the private world of pastoral signals a startling confrontation between the antithetical modes of agency and organization suggested throughout Harvey's treatises.[2] It is doubtful that Marvell's "Nymph Complaining," even though composed after the publication of Harvey's *Circulatio*, is alluding directly to this critical moment in Harvey's treatise. But the inarguable associations of Harvey's doe and Marvell's fawn with the sacrificial monarch draw these two texts, willy-nilly, into the same figurative orbit of regicidal allegory. This conjunction of cervine images bespeaks as well another conceptual realm shared by these two writers. To be sure, Harvey chose to vivisect deer for the primary reason of their availability: "by the favour and bounty of my Royal Master," he later reflected, "I had great store of his Deere at my devotion . . . and license to dissect and search into them."[3] But deer bore in the seventeenth century a number of loaded associations that Harvey, no less than Marvell, would have had difficulty overlooking. As Lyndy Abraham has persuasively demonstrated, the image of the deer—either doe or fawn—was deeply inscribed in the symbology of mid-seventeenth-century alchemy. The deer, as *cervus fugitivus*, the fleeing hart, was deployed by speculative alchemists to figure forth the mercurial principle of alchemy itself.[4] Harvey's doe and Marvell's fawn represent both monarch and the potentially antimonarchic principle of

2. The Civil War specificity of the "wanton Troopers" has been established by Edward S. Le Comte, who notes that the word *troopers* entered the language in 1640 as a term for the Parliamentary army, in "Marvell's 'The Nymph Complaining for the Death of Her Faun,'" *Modern Philology* 50 (1952): 100.

3. Harvey, *Anatomical Exercitations* (1653), p. 397.

4. Abraham, *Marvell and Alchemy*, pp. 248–54.

alchemical vitalism, functioning thereby as careful embodiments of the impossible dialectic at the heart of the Vitalist Moment.

The Royalist Harvey no doubt bewailed the death in 1649 of his "Royal Master." But we can extract from Harvey's account of this vivisection a theoretical faith in the organizational compensation for the demise of heart, hart, and king: there survives the loss of these centralizing forces a meaningful, productive agent in the continuously laboring blood. Andrew Marvell, of course, is never so sanguine about sanguineous efflux. And I wish to press the tenuous link between treatise and lyric in order to test the suggestion in Harvey—implicitly, in any of the vitalists—of vitalism's triumph. Marvell's poem can be seen to supply an ironic, explicitly tragic reading of the experimental discovery in Harvey's *Circulatio*. Whereas Harvey's dissection of deer reveals the fruitful labor of blood, Marvell's Nymph dismisses the compensatory or redemptive consequences of her deer's death:

> Though they should wash their guilty hands
> In this warm life-blood, which doth part
> From thine, and wound me to the Heart,
> Yet could they not be clean: their Stain
> Is dy'd in such a Purple Grain.
> There is not such another in
> The World, to offer for their Sin.
>
> (18–24)

As the Nymph's discourse of blood-guilt makes clear, the wanton troopers have modeled their murder of the fawn rather optimistically on the premier cultural topoi of redemptive sacrifice: the murder of Christ or king, of either of whom it can be said, "There is not such another in / The World." But the Nymph hastens to interpret for us the cruel irony that inevitably attends such sacrificial actions: the troopers have destroyed in this imitative sacrifice not only their king, whose royal purple stains their hands, but the very emblem of vitalist agency—the fawn's "warm life-blood"—which might reasonably be thought to found the principle of independent action in a post-monarchic world.

Before its random collision with Civil War gunshot, Marvell's fawn, we are told, was poised to enjoy a transformation that bore all the markings of a vitalist fantasy:

> But all its chief delight was still
> On Roses thus its self to fill:
> And its pure virgin Limbs to fold

In whitest sheets of Lillies cold.
Had it liv'd long, it would have been
Lillies without, Roses within.
                    (87–92)

With the surprising eloquence that so often attends this poet's retrospective anticipations of a lost vitalist possibility, the Nymph imagines what could have been, "Had it liv'd long," the vitalist transformation of the fawn's body. In figuring this bodily revolution that never happened, Marvell displaces the divinely guided Ovidian metamorphosis onto the more naturalistic terrain of material causation, depicting the concrescence of flora and fauna as a digestive product of self-moving matter. Milton's Raphael will describe for the unfallen Adam and Eve the mysterious process by which their bodies, in metabolizing their "corporal nutriments," "may at last turn all to Spirit, / Improv'd by tract of time, and wing'd ascend / Ethereal" (5.496–99).[5] Like Raphael, who will make this monistic progress contingent on their maintenance of virtue, the Nymph suggests here that the fawn will metamorphose on the moral strength of its "pure virgin Limbs": impelled by neither a divine will nor external pressure, the vital body fulfills its potential by an internally generated process of progressive bodily purification.

We have observed throughout this book the pattern by which the foundations of liberal agency must be laid on the ruins of its authoritarian predecessor. But Marvell's poem here makes especially clear the even greater complication that can beset the poetics of vitalist agency. It is not simply monarchy but vitalism itself—the principle of autonomous agency symbolically embodied in the "life-blood" of the fawn—that is sacrificed at the fawn's death. The "round stream" of blood breaking forth from Harvey's doe was the visible emblem of the vital powers infused throughout the liberally organized animal body. But, as Marvell's Nymph helps us see, the power of matter with which that liberal body is infused can be named and identified only as it is drained from view and rendered, we must suppose, lifeless and inert. In figuring this ironic pattern in explicitly elegiac terms, Marvell formalizes the consciousness of the lost promise of vitalist doctrine that constitutes the suppressed burden of all the texts we have examined. Vitalism is sacrificed almost at the moment of its inception.

Here we can turn to the second of the two elegiac topoi with which the vitalists mourn the unfulfilled, unfulfillable promise of their revolutionary materialist philosophy. After figuring the sacrifice of the liberatory potential

5. Danielson discusses the contemporary prevalence of this belief in unfallen perfectibility in *Milton's Good God*, pp. 210–14.

of animal vitality, Marvell's elegy for the Vitalist Moment presents itself once again as exemplary, as it concludes with an expression of what is unquestionably this period's chief literary emblem of vitalism's failure: the degenerative transformation of body into stone. The final image of "The Nymph Complaining," the peculiar figure of the hardening of malleable body, crystallized not only for Marvell but, as we will see, for a number of midcentury intellectuals the fate awaiting the vitalist ideal. Having witnessed the devitalization of her beloved pet, the animate Nymph looks ahead to a final translation to inanimate matter:

> First my unhappy Statue shall
> Be cut in Marble; and withal,
> Let it be weeping too: but there
> Th' Engraver sure his Art may spare;
> For I so truly thee bemoane,
> That I shall weep though I be Stone:
> Until my Tears, still dropping, wear
> My breast, themselves engraving there.
> There at my feet shalt thou be laid,
> Of purest Alabaster made.
>
> (111–20)

Imagining herself an "unhappy Statue . . . cut in Marble," her fawn, "Of purest Alabaster made," the Nymph seeks elegiac commemoration in the art of the sculptor. As so often, Marvell worries the justice of an artificer's action: when, and when not, "Th' Engraver sure his Art may spare." But the concern here with the proper representation of flesh also reflects the abiding interest in the agential and organizational implications of fleshly constitution. Vitalism's defeat at the muscled hand of mechanism entails, by means of the organizational analogy whose pressure we have felt throughout this book, a range of political implications. And Marvell at the end of "The Nymph Complaining" pursues those implications with a reference to the foundational text of mechanism, Descartes's *Discourse on Method*, which made its appearance in English in the fateful year 1649.

The gesture in the *Discourse* that most shocked English intellectuals was Descartes's identification of animals not as vital beings but as mechanical assemblages of parts.[6] Human beings, of course, were in possession of the

6. René Descartes, *Discourse on Method and Meditations on First Philosophy*, trans. Donald A. Cress (Indianapolis: Hackett, 1980), p. 30. See Marjorie Hope Nicolson, "Early Stage of Cartesianism in England," *Studies in Philosophy* 26 (1929): 451–74; Sterling Lamprecht, "Role of Descartes in Seventeenth-Century England," *Studies in the History of Ideas* 3 (1935):

"immaterial substance" of the soul, which liberated them at least in part from the mechanist's prison house of automation. But with the assumption "that lack of freedom was inherent in all mechanical devices," Descartes could insist that animals, which possessed no such soul, were but rank and lifeless matter ingeniously pieced together by the hand of God; they were no freer, according to Descartes, than "moving statues."[7] Ralph Cudworth, though no friend of vitalism, was horrified that Descartes could imagine a "Dead and Wooden World, as it were a Carved Statue, that hath nothing neither *Vital* nor *Magical* at all in it."[8] Henry More, too, turned "with abhorrence from . . . the sharp and cruel blade" of Cartesian mechanism, which "dared to despoil of life and sense practically the whole race of animals, metamorphosing them into marble statues and machines."[9] The emotional charge behind these reactions to the mechanist devitalization of animals is, I think, understandable. The stripping of life and freedom from the least of God's creatures was easily experienced as a prelude to an attack on the life and freedom of the greatest. And in the Nymph's complex response to the death of her fawn, we find her resignation to a body of thought that had already killed—at least theoretically—the entire animal kingdom, and might well begin to destroy its human counterpart.

The elegy for the liberty granted matter by the vitalist vision is always, at least in part, an elegy for the political liberty for which vitalism could serve as ontological ground. The political implications of vitalism's defeat, and of the figuration of that defeat as the transmutation of body to stone, are perhaps nowhere so pronounced as in the elegy for vitalism that Milton includes in *Paradise Lost*. We saw in Chapter 4 how Milton invested his own relation to the political excitement of the Vitalist Moment in his many representations of the chaotic abyss. The almost uniformly animated abyss of *Christian Doctrine* and Book Three of *Paradise Lost* bespoke, I suggested, the most egalitarian of Milton's political impulses. But the hierarchical monism of Books Five and Seven, pushing in its most desperate expressions toward the dualist segregation of animate matter from the tartareous dregs of the abyss, seemed to embody Milton's later resignation to the Council of State's authoritarian

---

184–240; and L. D. Cohen, "Descartes and Henry More on the Beast Machine," *Annals of Science* 1 (1936): 48–61.

7. Mayr, *Authority, Liberty, and Automatic Machinery*, p. 66. In *Milton among the Philosophers*, pp. 208–9, Fallon discusses a related image, the figure of the congelation of spirit into matter, in the work of Milton and Anne Conway.

8. Cudworth, *The True Intellectual System of the Universe* (I.iii. sec. XXXVII), in *Cambridge Platonists*, ed. Patrides, p. 290.

9. Letter to Descartes, dated December 11, 1648; quoted in Cohen, "Descartes and Henry More," p. 50. Cf. Margaret Cavendish, who discusses "animal bodies, whose parts," she insists, "all have animal life" (*Observations upon Experimental Philosophy*, p. 47).

hold on an unruly Commonwealth. Between those two conflicting sets of material representation, the death of at least a portion of the animate cosmos can be deemed to have occurred. And Milton, I think, mourns that death in the final and perhaps most shocking representation of chaos in all of *Paradise Lost*. He names and responds to the death of vitalism in his reintroduction of that allegorical grotesque, Death himself.

The outsized allegorization of death reappears at this point in Milton's poem as the fallen explanation for the new mortality of Adam and Eve, who, according to a less allegorical, natural philosophical explanation, had not known, in eating the fruit, that they were "eating Death" (9.792). The passage by which the allegorical characters Sin and Death will now travel to Earth is

> a Bridge
> Of length prodigious joining to the Wall
> Immoveable of this now fenceless World
> Forfeit to Death; from hence a passage broad,
> Smooth, easy, inoffensive down to Hell.
> (10.301–5)

The substance from which this bridge is constructed is, of course, the ideologically conflicted matter of chaos. The abyss Milton represents in this scene bears little resemblance to the vital matter from which the Father created the world; Milton has revived instead his depiction of the anarchic, specifically Hobbesian chaos that Satan traversed at the end of Book Two. If the Father's brooding impregnation of the abyss had generated for Milton elsewhere in the poem the ontological origin of the Vitalist Moment, then the vulgar imitation of that act performed by Sin and Death lays the foundation for a far less promising age of mechanical reproduction. Their construction of the bridge from Hell occurs in two stages, each represented by a central topos from vitalism's debate with mechanism.

The first stage of this public works project involves the forceful aggregation of inert particles. In defending his vitalist thesis of spontaneous generation against its many opponents, William Harvey attacked those materialists who imagined creation as a mechanical aggregation of atoms or elements, "as if (forsooth) *Generation* were nothing in the world, but a meer Separation, or Collection, or Order of things."[10] Although Sin and Death still inhabit Milton's putatively vitalist chaos, they oversee precisely the type of

---

10. Harvey, *Anatomical Exercitations*, p. 51.

conglomeratic generation Harvey dismissed as nonvitalist and therefore impossible:

> Then Both from out Hell Gates into the waste
> Wide Anarchy of *Chaos* damp and dark
> Flew diverse, and with Power (thir Power was great)
> Hovering upon the Waters; what they met
> Solid or slimy, as in raging Sea
> Tost up and down, together crowded drove
> From each side shoaling towards the mouth of Hell.
>   (10.282–88)

Earlier in the poem, Milton had asserted the generative power of the vital virtue immanent within chaotic matter. But the only power he acknowledges here is the mechanical strength these immaterial beings impose on chaos from outside. Their power *over* matter, quite different from the power *of* matter celebrated by the vitalists, is so divorced from the substance of chaos itself that it requires of Milton an awkward parenthetical recognition of its isolated majesty: "(thir Power was great)."

The second stage of the bridge's construction returns us to the topos of petrifaction introduced by Descartes and derided by his natural philosophical opponents. The constructive actions here—the crowding together and shoaling up of the random parts of chaos—assume a natural philosophy in no way conformable to the monistic materialism Milton laid out in *Christian Doctrine* or *Paradise Lost*. The vital body of chaos, which had provided the basis for Milton's theodicy and the intellectual underpinnings of his most liberal social and political principles, is, quite simply, dead. The poem marks the demise of its vitalism much as Marvell had in the "Nymph Complaining," with the loaded image of the transfiguration of body into stone:

> The aggregated Soil
> Death with his Mace petrific, cold and dry,
> As with a Trident smote, and fix't as firm
> As *Delos* floating once; the rest his look
> Bound with *Gorgonian* rigor not to move.
>   (10.293–97)

This active transformation of flexible liberal matter into adamant enacts on the level of narrative the oppressive intellectual process Milton had witnessed in the 1650s, the new philosophical conceptualization of all body, including flesh, *as* adamant. Cementing the fact of vitalism's death, Milton has Death bind matter with his "look," his debased natural philosophical vision, forcing it thereby "not to move." In a single glance, the once self-moving matter of chaos is reduced to the "black tartareous cold Infernal dregs / Adverse to life" that had even in Book Seven occupied no more than a discrete, purgeable portion of the Miltonic abyss.

Milton moves quickly to supply the political analogy of this scene's strange figuration of the mechanist ontology: "So, if great things to small may be compar'd," he tells us, then Death's tartareous passage may be likened to the bridge by which the Persian king Xerxes tied Europe to Asia, "the Liberty of *Greece* to yoke" (10.306–7). As we have seen, the new philosophy of spiritless matter, un-self-moving, driven solely by impact, seemed perhaps too compelling a rationalization for the theories of conquest and justifiable force that had been gathering strength since the onset of civil war. The forms of philosophical determinism that proceeded logically from such mechanistic ontologies worked easily to bolster a politics of force and centralized government. David Quint has unpacked the political implications of Milton's classical imagery here. Identifying Death's fixation of the floating island of Delos as Milton's parody of a commonplace Royalist image for the Restoration, Quint argues persuasively that the comparison of Death with Xerxes is Milton's declamation of a "similar enslavement of England's free Commonwealth to the royal tyranny of Charles II."[11] The petrifying constraint of the liberal matter of chaos indeed encodes an onset of authoritarianism of which the Stuart Restoration was for Milton a type. But if this scene constitutes, as I am suggesting, Milton's elegy for a vitalist philosophy of agency and organization, then we can take Quint's observations even further. The unregenerate exercise of "Power" figured here might bespeak *any* justification of authoritarian control; and we are obliged, I think, to include in this category Milton's own late proposal for the military empowerment of the Puritan Council of State (their power would be great).[12] Once a self-proclaimed "son of the soil," John Milton had, by the time he wrote *Paradise Lost*, smitten the generative soil of vitalism "As with a Trident." Reconstituting a once vital earth into the fitter soil of devitalized matter, Milton distanced himself from the most liberal strains of his own natural and political philosophy. In this startling depiction of an all too ready and easy path from Hell to earth, Milton laments not only the death of the Vitalist Moment or the evil days of monarchic resurgence. He laments, more darkly, the hardening of his own political sensibility.

I believe there was some justice in my initial historical identification of a

11. Quint, *Epic and Empire*, pp. 271, 270.
12. See *Works of Milton*, 6:141.

"Vitalist Moment," a concentrated period of radical intellectual activity that punctuated the middle of the seventeenth century. It may be desirable now, though, in light of this book's literary focus, to extend the term beyond its use as a historical marker of that brief period, from 1649 to 1652, in which vitalism acquired its early formal theorizations. The most characteristic feature of all the century's expressions of animist materialism, including those Restoration revivals appearing well after the peak of vitalist excitement, is the curiously conflicted literary manner in which the principle of self-moving matter is so often asserted. There are "vitalist moments" identifiable throughout the literature as specifically textual events. These vitalist moments, indicated here by the lower case, are those brief and overdetermined instances in treatise or poem in which the ideological promise of the vitalist ontology is revealed for its fragility; they are the disruptive points at which the literary assertion of vitalism's organizational potential lapses into a lament for its defeat. We have experienced these rhetorical moments several times: in Marvell's representation of the Mower's fall and Mary Fairfax's future marriage; in Milton's discovery of tartar in an otherwise animate chaos, and his sudden shift from a vitalist to a voluntarist account of the expulsion; and in the awkward gestures toward external motivation in Cavendish's otherwise internalist science of self-moving matter. I want now to reflect on the possible precipitants of these rhetorically charged, thematically pivotal moments that enliven and disrupt the works of Harvey, Winstanley, Marvell, Milton, and Cavendish. These textual events, I think, can be usefully calibrated to a wider set of sociopolitical pressures bearing down in the years of the Vitalist Moment. Sensitive to the possible alliance of rhetorical effect and historical event, we can speculate on the reasons why it was not vitalism, or any thoroughgoing philosophy of material self-motion, that ultimately subtended the agential and organizational cluster of principles we recognize as liberalism.

We have in Cavendish's *Observations* the most self-conscious narrativization of the momentous crises of conflicted vitalist allegiance that constitute this culture's vitalist moments. Nearly all the monistic materialists studied here—with the exception perhaps of Winstanley—evince a certain discomfort in the face of the notion of the uniform, or "egalitarian," diffusion of the power of animation. In a preface to her *Observations upon Experimental Philosophy*, Margaret Cavendish relates a moment of intellectual crisis at which she entertained the "New Thought" that matter might not be hierarchically divided into animate and inanimate portions but might be, as Winstanley had thought, entirely and unequivocally animated: these "New Thoughts endeavouring to oppose and call in question the Truth of my former Conceptions, caused a war in my mind, which in time grew to that height, that they were hardly able to compose the differences between themselves."[13] This war in her mind was won, Cavendish assures us, by the "former thoughts" of the hierarchical ordering of animate over inanimate matter: the "New Thoughts" of an absolute egalitarianism were successfully suppressed. Otherwise, as she argues in a slightly different context later in the book, if all parts of matter were equally possessed of knowledge and reason, "they would all be Governours, but none would be governed."[14] When we consider the extent to which agential and organizational propositions in natural philosophy seemed inevitably to prompt corresponding propositions within social and political spheres, it only stands to reason that the theorization of the politics of matter reproduces the conflicts besetting the prevailing theorization of the politics of the society at large. It should come as no surprise that the tension between competing natural philosophies presented itself to Cavendish as a military confrontation, holding as a consequence nothing less than the question of government.

A Royalist such as Cavendish or Harvey would naturally resist a thorough unfolding of a vitalist philosophy when its political implications became too difficult to ignore. But what needs now to be explained is the reason for which the *non*-Royalists examined here exhibit such discomfort with the organizational consequences of vitalism. Cavendish's image of a civil war taking place in the mind might best reflect not the conflicts between Royalists and Parliamentarians but the much knottier tension, in the 1650s, between republican Independents and the radical sectarians. Christopher Hill has described the revolution in mid-seventeenth-century England that "never happened."[15] At the very moment Independent Puritans were successfully shifting political authority from a monarch and an aristocracy to men of property, another, far more radical revolution threatened. The Levellers, for example, emerging from the Independent movement as an identifiable group in the later 1640s, sought to intensify the democratization of the polity, church, and courts. Soon the Diggers would go even further. Calling not merely for expansion of the franchise or the limitation of monopolies, the Diggers pushed, much more dangerously, for the communalization of private property. The fact that this radical revolution never happened, of course, was not the work of the disempowered Royalists. The suppression of this other revolution was the work, rather, of the propertied leaders of the Com-

---

13. Cavendish, *Observations upon Experimental Philosophy*, sig. h1[h]^r.

14. Ibid., p. 142.

15. Hill, *World Turned Upside Down*, p. 15. Nancy Armstrong and Leonard Tennenhouse discuss this attitude toward the failed egalitarian revolution as an effect of twentieth-century historiographical predisposition, in *The Imaginary Puritan: Literature, Intellectual Labor, and the Origins of Personal Life* (Berkeley: University of California Press, 1992), pp. 52–53.

monwealth, invested as they were in consolidating political advantage, cementing political gains.[16] The culture's most radical factions no doubt
experienced a symbolic empowerment in the execution of the king in January
1649. But the excitement for these groups would last only for a moment. It
was in May of the same year that mutinous Levellers in the army were
forcibly quelled at Burford. "One set of godly republicans," as Nigel Smith
has written, "trounced another set in the name of maintaining stability in
the early days of the new state."[17] The army-led suppression of the far more
radical Diggers would follow in due course, the routing of that movement
being successfully completed by 1651. The most conceptually daring elements of the Revolution had been defeated long before the restoration of
monarchy.

Hill and Pocock have described the crucial conflict during the 1650s between the republican commitment to an oligarchy of "saints" and the more
radical empowerment of the "people."[18] If we are to ascribe a political impetus to the figurations of defeat discernible even within the earliest assertions of vitalism, it would be the move to suppress the radical populists by
the newly entrenched proponents of republican stability. For a godly anti-
monarchist such as Milton, the saintly oligarchy making up the proposed
Council of State could not depend on the people to free themselves from
the authoritative paradigm of monarchy. The poet was drawn in his late
political writings to voice the paradoxical desire to secure an abstract principle of individual freedom through the military suppression of recalcitrant
individuals. A republican such as Marvell or Fairfax, we have seen, could not
permit the new distribution of political authority to extend to those not in
possession of land. For each of these figures, the suspicion of the general
unfitness of the "people" rendered an egalitarian polity unfeasible, even
frightening. And in the absence of a commitment to an egalitarian commonwealth, the once promising ontology of freedom and equality—the science
of vitalism—quickly proved too threatening. Depending for its survival on
the suppression of political energies that endangered its stability, an orderly
regime could not afford to entertain a metaphysics whose insistence on God's
equitable diffusion of spirit in matter could be so easily construed to guarantee an egalitarian politics.

Historians have recently brought to light the extent to which radical po-

16. See the study of the defeat of the Levellers and Diggers in Hill's *Experience of Defeat*,
pp. 29–50.
17. Nigel Smith, *Literature and Revolution in England, 1640–1660* (New Haven: Yale University Press, 1994), p. 148.
18. See Hill, *Experience of Defeat*, pp. 288–90; and Pocock, *Machiavellian Moment*, pp. 361–
400.

litical sentiment survived the Restoration.[19] Vitalism, too, had its scattered proponents after 1660. The radical Protestant and political Independent Henry Stubbe would assert a vitalist ontology well into the 1670s.[20] John Locke would offer a brief, perhaps halfhearted attempt to revitalize the earlier tradition with a remark in his *Essay concerning Human Understanding* that aroused emotionally charged attacks throughout the next century.[21] And various forms of the vitalist tradition would be kept alive in a marginal eighteenth-century movement identified by one historian as the "radical enlightenment," a phenomenon with a demonstrable origin in the vitalist writers of the English Revolution.[22] It should not be thought, however, that monistic materialism would ever regain the excitement or authority it enjoyed for the brief moment in the middle of the seventeenth century. The fact that one of the most formal and thoroughgoing articulations of a vitalist world-view, Francis Glisson's late *Tractatus de natura substantiae energetica* (1672), remains untranslated to this day attests to the vitalists' inability, or unwillingness, after the Vitalist Moment to reassert this ontology's place in the popular imagination. The political defeat of the sectarians, with whom, I believe, vitalism would on some level forever be aligned, was overwhelming. The good cause of vitalism was necessarily now the good *old* cause, too good perhaps to be fought for; a radical animism could persist into the Restoration as little more than an imperilled vision of matter, blurred and fleeting.

A philosophy of independent agency and decentralized organization would survive the death of the Vitalist Moment. But the "liberal" ideology that would begin to establish itself in the later part of the century had for the most part divorced itself from the vitalist ontology with which the principles of egalitarian association and free action had been linked at midcentury. We have seen suggested in the vitalist moments marking all the texts we have examined the reasons why this divorce was inevitable. The monistic union of matter with spirit provided all these writers with a powerful and attractive explanation of movement, process, and change. But any formulation of the infusion of matter with spirit continually opened up the possibility, as it did for Gerrard Winstanley, of the *uniform* infusion of matter with spirit. And this thesis of the uniform spiritualization of the soil of creation prompted a thesis, as it certainly had for Milton in 1651, of the uniform spirit and reason

19. See, for example, Greaves, *Deliver Us from Evil.*
20. See J. R. Jacob, *Henry Stubbe.*
21. See Yolton, *Thinking Matter*, p. 17; and above, Chapter 1, note 27.
22. M. C. Jacob, *Radical Enlightenment.* See also P. M. Heimann, "Voluntarism and Immanence: Conceptions of Nature in Eighteenth-Century Thought," *Journal of the History of Ideas* 39 (1978): 271–83.

of the "sons of the soil" (7:33). The philosophy of vitalism subtended assumptions of agency and organization too dangerously egalitarian for the republican liberals of the later period.

The principles of social and political independence that asserted themselves as liberal by the end of the century emerged, I think, from a body of thought that pointedly *de*vitalized the body of creation. Liberalism would reject the vital matter of revolution when it shifted its conceptual basis in the ontology of monism for a more politically functional origin in the economics of contract. The new member of the liberal polity would be the negotiating citizen rather than the divinized saint. His freedom would be figured as contractual consent rather than choice, his citizenship founded on a principle of obligation rather than an innate capacity for virtue or reason. If there is a contemporary social and political movement with which the fate of the Vitalist Moment was most closely tied, it would be the Engagement controversy, which teased the consciences of both Royalists and Parliamentarians from 1649 through the 1650s.[23] Republican Independents, impelled increasingly to identify the citizen in terms of his obedience to a de facto government, began at this point the difficult disengagement of republican ideology from the libertarian and antinomian philosophies that had fomented such excitement in the late 1640s.

There was a point in the Vitalist Moment, as I argued at the end of Chapter 1, at which the etymological root of the English *individual* might have been grounded indisputably in the vitalist lexicon: the monistic doctrine of the body's indivisibility from the soul—literally, its "individuality"—was in a position to supply the ontological ground for the free agency of the new, self-authorizing member of the polity known as the "individual." But, as we have seen repeatedly, the vitalist ontology of liberalism was asserted, at best, in fits and starts. In the wake of vitalism's dissipation—a process almost contemporaneous with vitalism's emergence—the English *individual* was obliged to emerge from the linguistic field of a competing materialism, that of mechanism. The individual arose, then, not as the self-moving body indivisibly aligned with soul, but as a personification of the *individuus*, the atomic particle that served as the indivisible building block of the authoritative edifice of mechanism. We have seen throughout this book the felt implications of this fall from vitalism into mechanism. The new, "atomic"

23. In *Destiny His Choice: The Loyalism of Andrew Marvell* (Cambridge: Cambridge University Press, 1968), one of the best discussions of the Engagement controversy, John M. Wallace rightly situates Marvell in the contemporary debates concerning loyal engagement. But depicting Marvell as a more or less unambivalent Engager, Wallace fails to see how the pastoral lyrics lament the fall of political identity from its putative origins in natural authority to its devitalized domicile in civil engagement.

individual must establish himself or herself as a figure whose free agency has been subjected to the authority of a higher power and whose political identity is negotiated in terms of an allegiance to that power. Expelled from an origin in vitalism into a Hobbesian ontology of realpolitik, the individual falls from the status of freely willing saint to that of citizen, contracting with and then obeying the appointed power. We have in Marvell's Complaining Nymph a moving response to this devitalization of the vital political animal into the new adamant liberal. Looking back at the radical bodily transformation that did not take place, looking ahead to the hardened matter from which all things will soon be cut, Marvell's Nymph succumbs to perpetual mourning.

BIBLIOGRAPHY

Primary Sources

Agrippa, Henry Cornelius. *Three Books of Occult Philosophy*. Trans. J[ohn]. F[rench]. London, 1651.

Aquinas, Thomas. *The Summa Theologica of St. Thomas Aquinas*. Vol. 3. *Supplement*. New York: Benziger Brothers, 1948.

Aristotle. *The Politics of Aristotle*. Ed. and trans. Ernest Barker. Oxford: Clarendon, 1946.

Arminius, James. *The Writings of James Arminius*. Trans. James Nichols and William R. Bagnall. 2 vols. Grand Rapids, Mich.: Baker Book House, 1956.

Astell, Mary. *A Serious Proposal to the Ladies for the Advancement of their True and Greatest Interest*. London, 1696.

Augustine. *The City of God*. Trans. Henry Bettenson. Harmondsworth: Penguin, 1984.

Bacon, Francis. *The Advancement of Learning and New Atlantis*. London: Oxford University Press, 1906.

——. *Works*. 15 vols. Ed. James Spedding, R. L. Ellis, and D. D. Heath. Boston: Houghton Mifflin, 1861.

Baxter, Richard. *Christian Directory*. London, 1673.

Boehme, Jakob. *The epistles of Jacob Behmen. Very usefull and necessary for those that read his writings*. Trans. John Ellistone. London, 1649.

Boyle, Robert. *Occasional Reflections upon Several Subjects*. London, 1665.

——. *Selected Philosophical Papers of Robert Boyle*. Ed. M. A. Stewart. New York: Barnes and Noble, 1979.

——. *The Works of the Honourable Robert Boyle*. Ed. Thomas Birch. 2d ed. 6 vols. London: J. and F. Rivington, 1772.

Browne, Thomas. *The Major Works*. Ed. C. A. Patrides. Harmondsworth: Penguin, 1977.

Calvin, John [Jean]. *Commentaries on the Epistle of Paul the Apostle to the Romans.* Trans. J. Owen. Edinburgh: Calvin Translation Society, 1849.

——. *Institutes of the Christian Religion.* Trans. John Allen. 2 vols. Philadelphia: Westminster Press, 1935.

Caussin, Nicholas. *The Holy Court in Five Tomes.* Trans. T[homas]. H[awkins]. London, 1650.

Cavendish, Margaret. *The Blazing World and Other Writings.* Ed. Kate Lilley. Harmondsworth: Penguin, 1992.

——. *Life of the Thrice Noble . . . William Cavendishe.* London, 1667.

——. *Natures Picture Drawn by Fancies Pencil to the Life.* 2d ed. London, 1671.

——. *Observations upon Experimental Philosophy. To Which is added The Description of a New Blazing World.* London, 1666.

——. *Orations of Divers Sorts.* London, 1662.

——. *Philosophical and Physical Opinions.* London, 1655.

——. *Philosophical and Physical Opinions.* 2d ed. London, 1663.

——. *Philosophicall Fancies.* London, 1653.

——. *Philosophical Letters: or, Modest Reflections upon some Opinions in Natural Philosophy By the Thrice Noble, Illustrious, and Excellent Princess, The Lady Marchioness of Newcastle.* London, 1664.

——. *Poems, and Fancies: Written by the Right Honourable, The Lady Newcastle.* London, 1653.

——. *CCXI Sociable Letters.* London, 1664.

——. *The Worlds Olio.* London, 1655.

Charleton, Walter. *The Darknes of Atheism Dispelled by the Light of Nature: A Physico-Theological Treatise.* London, 1652.

——. *Natural History of Nutrition, Life, and Voluntary Motion.* London, 1659.

——. *Physiologica Epicuro-Gassendo-Charltoniana: Or a Fabrick of Science Natural, Upon the Hypothesis of Atoms.* 1654. New York: Johnson Reprint, 1954.

A Collection of Letters, Poems, etc. written to . . . the Duke and Duchess of Newcastle. London, 1678.

Conway, Anne. *Conway Letters: The Correspondence of Anne, Viscountess Conway, Henry More, and Their Friends, 1642–1684.* Ed. Marjorie Hope Nicolson. New Haven: Yale University Press, 1930.

Crab, Roger. *The English Hermite and Dagons-Downfall.* 1655, 1657. Ed. Andrew Hopton. London: Aporia, 1990.

Cromwell, Oliver. *Writings and Speeches of Oliver Cromwell.* Ed. W. C. Abbott. 4 vols. Cambridge: Harvard University Press, 1947.

Cudworth, Ralph. *A Treatise Concerning Eternal and Immutable Morality.* 2 vols. London, 1731.

——. *The True Intellectual System of the Universe.* London, 1678.

Davenant, William. *Sir William Davenant's Gondibert.* Ed. David F. Gladish. Oxford: Clarendon, 1971.

Descartes, René. *Discourse on Method and Meditations on First Philosophy.* Trans. Donald A. Cress. Indianapolis: Hackett, 1980.

——. *The Philosophical Writings of Descartes.* Trans. John Cottingham, Robert

Stoothoff, and Dugald Murdoch. 2 vols. Cambridge: Cambridge University Press, 1984–85.

Dorn, Geraldus. *A chymicall dictionary: explaining hard places and vvords met withall in the writings of Paracelsus, and other obscure authours*. London, 1650.

Evelyn, John. *An Essay on the First Book of T. Lucretius Carus, "De Rerum Natura."* London, 1656.

———. *Sylva, or a Discourse of Forest Trees*. 1664. 2 vols. London: Doubleday, 1908.

Everard, John. *The Divine Pymander of Hermes Mercurius Trismegistus, In XVII. Books*. London, 1650.

Firth, C. H., ed. *Stuart Tracts, 1603–1693*. New York: Cooper Square, 1964.

Fludd, Robert. *Mosaicall Philosophy: Grounded upon the Essentiall Truth or Eternal Sapience*. London, 1659.

French, John. *A new light of Alchymie: taken out of the fountaine of nature, and manuall experience*. London: Richard Cotes, 1650.

Glisson, Francis. *De rachitude*. London, 1650.

———. *Doctrina de Circulatione Sanguinis Haud Immutat Antiquam Medendi Methodum*. In Jeffrey M. N. Boss, "*Doctrina de Circulatione Sanguinis Haud Immutat Antiquam Medendi Methodum*: An Unpublished Manuscript (1662) by Francis Glisson (1597–1677) on Implications of Harvey's Physiology," *Physis* 20 (1978): 309–36.

———. *From "Anatomia Hepatis" (The Anatomy of the Liver), 1654*. English Manuscripts of Francis Glisson (1). Ed. Andrew Cunningham. Cambridge, England: Wellcome Unit for the History of Medicine, 1993.

———. *Tractatus de natura substantiae energetica s. de vita naturae ejusque tribus primis facultatibus, i. perceptiva, ii. appetitiva, iii. motiva, naturalibus*. London, 1672.

———. *A Treatise of the Rickets, being a Diseas common to Children*. Trans. Phil. Armin. London, 1651.

Gregory of Nyssa. *Ascetical Works*. Trans. Virginia Callahan. Washington, D.C.: Catholic University of America Press, 1967.

Hall, Thomas. *Vindiciae Literarum*. London, 1655.

Harrington, James. *The Censure of the Rota upon Mr Miltons Book, Entituled, The Ready and Easie way to Establish a Free Common-wealth*. London, 1660.

———. *The Political Works of James Harrington*. Ed. J. G. A. Pocock. Cambridge: Cambridge University Press, 1977.

Harvey, William. *The Anatomical Exercises of Dr. William Harvey, De Motu Cordis 1628: De Circulatione Sanguinis 1649: The First English Text of 1653*. Ed. Geoffrey Keynes. London: Nonesuch, 1928.

———. *The Anatomical Exercises of Dr. William Harvey . . . with the Preface of Zachariah Wood, Physician of Rotterdam*. London, 1653.

———. *Anatomical Exercitations, concerning the Generation of Living Creatures: To which are added Particular Discourses, of Births, and of Conceptions, &c*. London, 1653.

———. *Disputations touching the Generation of Animals*. Trans. and ed. Gweneth Whitteridge. Oxford: Blackwell Scientific Publications, 1981.

Herrick, Robert. *The Complete Poetry of Robert Herrick*. Ed. J. Max Patrick. New York: Norton, 1968.

Hobbes, Thomas. *The English Works of Thomas Hobbes*. 11 vols. Ed. William Molesworth. London: Bohn, 1839.

——. *Leviathan*. Ed. C. B. Macpherson. Harmondsworth: Penguin, 1951.

Howarth, R. G., ed. *Minor Poets of the Seventeenth Century*. London: Dent, 1953.

Locke, John. *Essay concerning Human Understanding*. Ed. Peter H. Nidditch. Oxford: Clarendon, 1975.

Marvell, Andrew. *Complete Poetry*. Ed. George deF. Lord. London: Dent, 1984.

——. *The Poems and Letters of Andrew Marvell*. Ed. H. M. Margoliouth, rev. Pierre Legouis. 2 vols. Oxford: Clarendon, 1971.

——. *The Rehearsal Transpros'd and The Rehearsal Transpros'd. The Second Part*. Oxford: Oxford University Press, 1971.

Methodius. *The Symposium: A Treatise on Chastity*. Trans. Herbert Musurillo. London: Longmans, 1958.

Milton, John. *Complete Poems and Major Prose*. Ed. Merritt Y. Hughes. New York: Odyssey, 1957.

——. *Complete Prose Works of John Milton*. Ed. Don M. Wolfe. 8 vols. New Haven: Yale University Press, 1953–82.

——. *Political Writings*. Ed. Martin Dzelzainis. Trans. Claire Gruzelier. Cambridge: Cambridge University Press, 1991.

——. *The Works of John Milton*. 18 vols. Ed. Frank Patterson et al. New York: Columbia University Press, 1931–40.

Misselden, Edward. *The Circle of Commerce. Or the Balance of Trade*. London, 1623.

——. *Free Trade. or, the meanes to make trade florish*. London, 1622.

More, Henry. *An Antidote Against Atheism*. 2d ed. London, 1655.

——. *A Collection of Several Philosophical Writings*. London, 1662.

——. *Enthusiasmus triumphatus*. London, 1656.

——. *The Immortality of the Soul, So farre forth as it is demonstrable from the Knowledge of Nature and the Light of Reason*. London, 1659.

——. *Opera omnia, tum quae latine, tum quae anglice scripta sunt. . . .* 2 vols. London, 1679.

Mun, Thomas. *A discourse of trade, from England unto the East Indies*. London, 1621.

——. *England's treasure by forraign trade*. 1623. London, 1664.

——. *The petition and remonstrance of the governor and company of merchants of London trading to the East-Indies*. London, 1628.

Overton, Richard. *An Arrow Against All Tyrants and Tyrany, shot from the Prison of New-gate into the Prerogative Bowels of the Arbitrary House of Lords and all other Usurpers and Tyrants Whatsoever*. London, 1646.

——. *Mans Mortalitie*. 1644. Ed. Harold Fisch. Liverpool: Liverpool University Press, 1968.

Paracelsus [Theophrastus von Hohenheim]. *Paracelsus: Selected Writings*. Ed. Jolande Jacobi. Trans. Norbert Guterman. Princeton: Princeton University Press, 1951.

——. *Paracelsus His Archidoxis: Comprised in Ten Books*. London, 1660.

——. *Philosophy Reformed & Improved in Four Profound Tractates . . . Discovering the Wonderfull Mysteries of the Creation, by Paracelsus: Being His Philosophy to the Athenians*. Trans. H. Pinnell. London, 1657.

Paré, Ambroise. *The Workes of that Famous Chirurgian Ambrose Parey*. Trans. Thomas Johnson. London: Thomas Cotes, 1634.

Patrides, C. A., ed. *The Cambridge Platonists*. Cambridge: Cambridge University Press, 1969.

Pepys, Samuel. *The Diary of Samuel Pepys*. Ed. Henry B. Wheatley. 6 vols. New York: Harcourt, Brace, 1938.

Puttenham, George. *The Arte of English Poesie*. Ed. Gladys D. Willcock and Alice Walker. Cambridge: Cambridge University Press, 1936.

Ross, Alexander. *Arcana Microcosmi: or, The hid Secrets of Man's Body discovered; In an Anatomical Duel between Aristotle and Galen concerning the Parts thereof: As also, By a Discovery of the strange and marveilous Diseases, Symptomes & Accidents of Man's Body. With a Refutation of Doctor Brown's Vulgar Errors, The Lord Bacon's Natural History, and Doctor Harvy's Book De Generatione, Comenivs, and Others*. London, 1652.

Sandys, George. *Ovid's Metamorphosis Englished, Mythologized and Represented in Figures*. Ed. Karl T. Hulley and Stanley T. Vandersall. Lincoln: University of Nebraska Press, 1970.

Shakespeare, William. *The Riverside Shakespeare*. Boston: Houghton Mifflin, 1974.

Sidney, Sir Philip. *The Countess of Pembroke's Arcadia*. Ed. Maurice Evans. Harmondsworth: Penguin, 1977.

Smith, John. *Select Discourses*. London, 1660.

Spenser, Edmund. *Spenser: Poetical Works*. Ed. J. C. Smith and E. de Selincourt. London: Oxford University Press, 1912.

Stanley, Sir Thomas. *The History of Philosophy. Containing those on whom the Attribute of Wise was Conferred*. 3 vols. London, 1655–60.

Taylor, Jeremy. *Holy Living*. 1650. Ed. P. G. Stanwood. Oxford: Clarendon, 1989.

A Third Collection of . . . Poems, Satires, Songs, &c. against Popery and Tyranny. London, 1689.

van Helmont, Jean Baptiste [Jan Baptista]. *Oriatrike, or Physick Refined*. London, 1662.

——. *Ortus Medicinae*. London, 1648.

——. *Ternary of Paradoxes. Translated, Illustrated, and Ampliated by Walter Charleton*. London, 1650.

Winstanley, Gerrard. *Selected Writings*. Ed. Andrew Hopton. London: Aporia, 1989.

——. *The Works of Gerrard Winstanley, With an Appendix of Documents Relating to the Digger Movement*. Ed. George H. Sabine. Ithaca: Cornell University Press, 1941.

Secondary Sources

Abraham, Lyndy. *Marvell and Alchemy*. Hants, England: Scolar, 1990.

Adams, Robert M. "A Little Look into Chaos." In *Illustrious Evidence: Approaches to English Literature of the Early Seventeenth Century*, ed. Earl Miner, pp. 71–92. Berkeley: University of California Press, 1975.

Aers, David, and Bob Hodge. " 'Rational Burning': Milton on Sex and Marriage." In *Literature, Language, and Society in England, 1580–1680*, ed. David Aers, Bob Hodge, and Gunther Kress. Dublin: Gill and Macmillan, 1981.

Aers, David, and Mary Ann Radzinowicz, eds. *Paradise Lost: Books VII–VIII*. Cambridge: Cambridge University Press, 1974.

Allen, Don Cameron. *Image and Meaning: Metaphoric Traditions in Renaissance Poetry*. 2d ed. Baltimore: Johns Hopkins University Press, 1968.

Appleby, Joyce Oldham. *Economic Thought and Ideology in Seventeenth-Century England*. Princeton: Princeton University Press, 1978.

Armstrong, Nancy, and Leonard Tennenhouse. *The Imaginary Puritan: Literature, Intellectual Labor, and the Origins of Personal Life*. Berkeley: University of California Press, 1992.

Aylmer, G. E. "The Religion of Gerrard Winstanley." In *Radical Religion in the English Revolution*, ed. J. F. McGregor and B. G. Reay. Oxford: Oxford University Press, 1984.

Baker, Herschel. *The Race of Time: Three Lectures on Renaissance Historiography*. Toronto: University of Toronto Press, 1967.

——. *The Wars of Truth: Studies in the Decay of Christian Humanism in the Earlier Seventeenth Century*. Cambridge: Harvard University Press, 1952.

Barkan, Leonard. *Nature's Work of Art: The Human Body as Image of the World*. New Haven: Yale University Press, 1975.

Barker, Arthur. *Milton and the Puritan Dilemma, 1641–1660*. Toronto: University of Toronto Press, 1942.

Barnes, Barry, and Steven Shapin, eds. *Natural Order: Historical Studies of Scientific Culture*. Beverly Hills: Sage, 1979.

Bedford, R. D. "Time, Freedom, and Foreknowledge in *Paradise Lost*." *Milton Studies* 16 (1982): 61–76.

Bennett, Joan S. *Reviving Liberty: Radical Christian Humanism in Milton's Great Poems*. Cambridge: Harvard University Press, 1989.

——. "Virgin Nature in *Comus*." *Milton Studies* 23 (1987): 21–32.

Berens, Lewis H. *The Digger Movement in the Days of the Commonwealth*. London: Simpkin, Marshall, Hamilton, Kent, and Co., 1906.

Berger, Harry, Jr. "Marvell's 'Garden': Still Another Interpretation." *Modern Language Quarterly* 28 (1967): 285–304.

——. "Marvell's 'Upon Appleton House': An Interpretation." *Southern Review* (Australia) 1, no. 4 (1965): 7–32. Reprinted in *John Donne and the Seventeenth-Century Metaphysical Poets*, ed. Harold Bloom. New York: Chelsea House, 1986.

Blamires, Harry. *Milton's Creation: A Guide through "Paradise Lost."* London: Methuen, 1971.

Bloch, Marc. *The Royal Touch: Sacred Monarchy and Scrofula in England and France*. Trans. J. E. Anderson. London: Routledge, 1973.

Bowerbank, Sylvia. "The Spider's Delight: Margaret Cavendish and the 'Female' Imagination." *English Literary Renaissance* 14 (1984): 392–408.

Brisman, Leslie. *Milton's Poetry of Choice and Its Romantic Heirs*. Ithaca: Cornell University Press, 1973.

Broadbent, John B. *Some Graver Subject: An Essay on "Paradise Lost."* London: Chatto and Windus, 1960.

Brown, Peter. *The Body and Society: Men, Women, and Sexual Renunciation in Early Christianity*. New York: Columbia University Press, 1988.

Burden, Dennis H. *The Logical Epic: A Study of the Argument of "Paradise Lost."* London: Routledge, 1967.

Burns, Norman T. *Christian Mortalism from Tyndale to Milton*. Cambridge: Harvard University Press, 1972.

Bylebyl, Jerome J., ed. *William Harvey and His Age: The Professional and Social Context of the Discovery of the Circulation*. Baltimore: Johns Hopkins University Press, 1979.

Carter, Richard B. *Descartes' Medical Philosophy: The Organic Solution to the Mind-Body Problem*. Baltimore: Johns Hopkins University Press, 1983.

Cassirer, Ernst. *The Individual and the Cosmos in Renaissance Philosophy*. Trans. Mario Domandi. Philadelphia: University of Pennsylvania Press, 1963.

Cattaneo, Mario A. "Hobbes's Theory of Punishment." In *Hobbes Studies*, ed. K. C. Brown, pp. 275–98. Cambridge: Harvard University Press, 1965.

Chambers, A. B. "Chaos in *Paradise Lost*." *Journal of the History of Ideas* 24 (1963): 55–84.

Chernaik, Warren L. *The Poet's Time: Politics and Religion in the Work of Andrew Marvell*. Cambridge: Cambridge University Press, 1983.

Christopher, Georgia B. *Milton and the Science of the Saints*. Princeton: Princeton University Press, 1982.

Clark, Elizabeth A. *Ascetic Piety and Women's Faith: Essays on Late Ancient Christianity*. Lewiston, New York: Edwin Mellen Press, 1986.

Clark, Mili N. "The Mechanics of Creation: Non-Contradiction and Natural Necessity in *Paradise Lost*." *English Literary Renaissance* 7 (1977): 207–42.

Clucas, Stephen. "Poetic Atomism in Seventeenth-Century England: Henry More, Thomas Traherne, and 'Scientific Imagination.'" *Renaissance Studies* 5, no. 3 (1991): 327–40.

Cohen, I. Bernard. "Harrington and Harvey: A Theory of the State Based on the New Physiology." *Journal of the History of Ideas* 55 (1994): 187–210.

Cohen, L. D. "Descartes and Henry More on the Beast Machine." *Annals of Science* 1 (1936): 48–61.

Colie, Rosalie L. *"My Ecchoing Song": Andrew Marvell's Poetry of Criticism*. Princeton: Princeton University Press, 1970.

———. *Paradoxia Epidemica: The Renaissance Tradition of Paradox*. Princeton: Princeton University Press, 1966.

Cooper, J. P. "Social and Economic Policies under the Commonwealth." In *The Interregnum: The Quest for Settlement, 1646–1660*, ed. by G. E. Aylmer. London: Macmillan, 1982.

Corns, Thomas N. *Uncloistered Virtue: English Political Literature, 1640–1660*. Oxford: Clarendon, 1992.

Cummings, Robert. "The Forest Sequence in Marvell's *Upon Appleton House*: The

Imaginative Contexts of a Poetic Episode." *Huntington Library Quarterly* 47 (1984): 179–210.

Curry, Walter Clyde. *Milton's Ontology, Cosmogony, and Physics*. Lexington: University Press of Kentucky, 1957.

Curtis, John G. *Harvey's Views on the Use of the Circulation of the Blood*. New York: Columbia University Press, 1915.

Danielson, Dennis. *Milton's Good God: A Study in Literary Theodicy*. Cambridge: Cambridge University Press, 1982.

Davies, Stevie. *The Idea of Woman in Renaissance Literature: The Feminine Reclaimed*. Brighton: Harvester, 1986.

——. *Images of Kingship in "Paradise Lost": Milton's Politics and Christian Liberty*. Columbia: University of Missouri Press, 1983.

Davis, Audrey B. *Circulation Physiology and Medical Chemistry in England, 1650–1680*. Lawrence, Kans.: Coronado, 1973.

Davis, J. C. "Gerrard Winstanley and the Restoration of True Magistracy." *Past and Present* 70 (1976): 78–92.

de Mause, Lloyd, ed. *The History of Childhood*. New York: Psychohistory Press, 1974.

Debus, Allen G. "The Chemical Debates of the Seventeenth Century: The Reaction to Robert Fludd and Jean Baptiste van Helmont." In *Reason, Experiment, and Mysticism in the Scientific Revolution*, ed. M. L. Righini Bonelli and William R. Shea, pp. 21–47. New York: Science History Publications, 1975.

——. *The Chemical Dream of the Renaissance*. Cambridge: Heffer, 1968.

——. *The Chemical Philosophy: Paracelsian Science and Medicine in the Sixteenth and Seventeenth Centuries*. 2 vols. New York: Science History Publications, 1977.

——. *The English Paracelsians*. London: Oldbourne, 1965.

——. *Science and Education in the Seventeenth Century: The Webster-Ward Debate*. London: Macdonald, 1970.

——, ed. *Science, Medicine, and Society in the Renaissance: Essays to Honor Walter Pagel*. 2 vols. New York: Watson, 1972.

Debus, Allen G., and Robert P. Multhauf. *Alchemy and Chemistry in the Seventeenth Century*. Los Angeles: William Andrews Clark Memorial Library, 1966.

Diamond, William Craig. "Natural Philosophy and Harrington's Political Thought." *Journal of the History of Philosophy* 16 (1978): 387–98.

Dijksterhuis, E. J. *The Mechanization of the World Picture: Pythagoras to Newton*. Princeton: Princeton University Press, 1986.

DiSalvo, Jackie. *War of Titans: Blake's Critique of Milton and the Politics of Religion*. Pittsburgh: University of Pittsburgh Press, 1983.

Dubrow, Heather. *A Happier Eden: The Politics of Marriage in the Stuart Epithalamium*. Ithaca: Cornell University Press, 1990.

Dunn, John. *Political Obligation in Its Historical Context: Essays in Political Theory*. Cambridge: Cambridge University Press, 1980.

Dunn, Kevin. "Milton among the Monopolists: *Areopagitica*, Intellectual Property, and the Hartlib Circle." In *Samuel Hartlib and Universal Reformation: Studies in Intellectual Communication*, ed. Mark Greengrass, Michael Leslie, and Timothy Raylor, pp. 177–92. Cambridge: Cambridge University Press, 1994.

——. *Pretexts of Authority: The Rhetoric of Authorship in the Renaissance Preface.* Stanford: Stanford University Press, 1994.

Easlee, Brian. *Witch-Hunting, Magic, and the New Philosophy: An Introduction to Debates of the Scientific Revolution, 1450–1750.* Sussex: Harvester, 1980.

Elmer, Peter. "Medicine, Religion, and the Puritan Revolution." In *The Medical Revolution of the Seventeenth Century,* ed. Roger French and Andrew Wear, pp. 10–45. Cambridge: Cambridge University Press, 1989.

Empson, William. *Milton's God.* London: Chatto and Windus, 1965.

——. *Some Versions of Pastoral.* New York: New Directions, 1950.

——. *Using Biography.* London: Chatto and Windus, 1984.

Enterline, Lynn. "The Mirror and the Snake: The Case of Marvell's 'Unfortunate Lover.'" *Critical Quarterly* 29, no. 4 (1987): 98–112.

Entzminger, Robert L. "Michael's Options and Milton's Poetry: *Paradise Lost* XI and XII." *English Literary Renaissance* 8 (1978): 197–211.

Erickson, Robert A. "William Harvey's *De motu cordis* and 'The Republic of Letters.'" In *Literature and Medicine during the Eighteenth Century,* ed. Marie Mulvey Roberts and Roy Porter, pp. 58–83. London: Routledge, 1993.

Ezell, Margaret J. M. *The Patriarch's Wife: Literary Evidence and the History of the Family.* Chapel Hill: University of North Carolina Press, 1987.

Fallon, Robert. *Milton in Government.* University Park: Pennsylvania State University Press, 1993.

Fallon, Stephen M. "The Metaphysics of Milton's Divorce Tracts." In *Politics, Poetics, and Hermeneutics in Milton's Prose,* ed. David A. Loewenstein and James Grantham Turner, pp. 69–83. Cambridge: Cambridge University Press, 1990.

——. *Milton among the Philosophers: Poetry and Materialism in Seventeenth-Century England.* Ithaca: Cornell University Press, 1991.

——. "'To Act or Not': Milton's Conception of Divine Freedom." *Journal of the History of Ideas* 49 (1988): 425–52.

Finlayson, Michael G. *Historians, Puritanism, and the English Revolution: The Religious Factor in English Politics before and after the Interregnum.* Toronto: University of Toronto Press, 1983.

Fish, Stanley. "Transmuting the Lump: *Paradise Lost*, 1942–1979." In his *Doing What Comes Naturally: Change, Rhetoric, and the Practice of Theory in Legal and Literary Studies,* pp. 247–93. Durham: Duke University Press, 1989.

Fitzmaurice, James. "Fancy and the Family: Self-Characterizations of Margaret Cavendish." *Huntington Library Quarterly* 53 (1990): 198–209.

Foucault, Michel. *Power/Knowledge.* Ed. Colin Gordon. Brighton: Harvester, 1980.

Fouke, Daniel C. "Mechanical and 'Organical' Models in Seventeenth-Century Explanations of Biological Reproduction." *Science in Context* 3 (1989): 365–81.

Fowler, Alastair. "Country House Poems: The Politics of a Genre." *Seventeenth Century* 1 (1986): 1–14.

——, ed. *John Milton: Paradise Lost.* Essex: Longman, 1971.

Frank, Joseph. *The Levellers: A History of the Writings of Three Seventeenth-Century Social Democrats.* Cambridge: Harvard University Press, 1955.

Frank, Robert G. *Harvey and the Oxford Physiologists: Scientific Ideas and Social Interaction.* Berkeley: University of California Press, 1980.

French, J. Milton, ed. *The Life Records of John Milton*. 5 vols. New York: Gordian Press, 1966.

French, Roger. "Harvey and Holland: Circulation and the Calvinists." In *The Medical Revolution of the Seventeenth Century*, ed. Roger French and Andrew Wear, pp. 46–86. Cambridge: Cambridge University Press, 1989.

Friedman, Donald. *Marvell's Pastoral Art*. Berkeley: University of California Press, 1970.

Froula, Christine. "When Eve Reads Milton: Undoing the Canonical Economy." *Critical Inquiry* 10 (1983): 321–47. Reprinted in *Canons*, ed. Robert von Hallberg, pp. 149–76. Chicago: University of Chicago Press, 1984.

Frye, Northrop. *The Return of Eden: Five Essays on Milton's Epics*. Toronto: University of Toronto Press, 1965.

Gallagher, Catherine. "Embracing the Absolute: The Politics of the Female Subject in Seventeenth-Century England." *Genders* 1 (1988): 24–39.

Geisst, Charles R. *The Political Thought of John Milton*. London: Macmillan, 1984.

Gelbart, Nina Rattner. "The Intellectual Development of Walter Charleton." *Ambix* 18 (1971): 149–68.

Gilbertson, Carol. "'Many Miltons in this One Man': Marvell's Mower Poems and *Paradise Lost*." *Milton Studies* 22 (1987): 151–72.

Gransden, K. W. "Time, Guilt, and Pleasure: A Note on Marvell's Nostalgia." *Ariel* 1, no. 2 (1970): 86.

Grant, Douglas. *Margaret the First: A Biography of Margaret Cavendish*. London: University of Toronto Press, 1957.

Greaves, Richard L. *Deliver Us from Evil: The Radical Underground in Britain, 1660–1663*. New York: Oxford University Press, 1986.

Greene, Thomas. *The Descent from Heaven: A Study in Epic Continuity*. New Haven: Yale University Press, 1963.

Grossman, Marshall. "Authoring the Boundary: Allegory, Irony, and the Rebus in *Upon Appleton House*." In *"The Muses Common-Weale": Poetry and Politics in the Seventeenth Century*, ed. Claude J. Summers and Ted-Larry Pebworth, pp. 191–206. Columbia: University of Missouri Press, 1988.

——. *"Authors to Themselves": Milton and the Revelation of History*. Cambridge: Cambridge University Press, 1987.

Guibbory, Achsah. *The Map of Time: Seventeenth-Century English Literature and Ideas of Pattern in History*. Urbana: University of Illinois Press, 1986.

Guillory, John. "Dalila's House: *Samson Agonistes* and the Sexual Division of Labor." In *Rewriting the Renaissance: The Discourses of Sexual Difference in Early Modern Europe*, ed. Margaret W. Ferguson, Maureen Quilligan, and Nancy J. Vickers, pp. 106–22. Chicago: University of Chicago Press, 1986.

——. "The Father's House: *Samson Agonistes* in Its Historical Moment." In *Remembering Milton: Essays on the Texts and Traditions*, ed. Mary Nyquist and Margaret W. Ferguson, pp. 148–76. New York: Methuen, 1987.

——. "From the Superfluous to the Supernumerary: Reading Gender into *Paradise Lost*." In *Soliciting Interpretation: Literary Theory and Seventeenth-Century English Poetry*, ed. Elizabeth D. Harvey and Katharine Eisaman Maus, pp. 68–88. Chicago: University of Chicago Press, 1990.

———. *Poetic Authority: Spenser, Milton, and Literary History*. New York: Columbia University Press, 1983.

Haber, Judith. *Pastoral and the Poetics of Self-Contradiction: Theocritus to Marvell*. Cambridge: Cambridge University Press, 1994.

Habermas, Jürgen. *The Structural Transformation of the Public Sphere: An Inquiry into a Category of Bourgeois Society*. Trans. Thomas Burger. Cambridge: MIT Press, 1991.

Hale, David George. *The Body Politic: A Political Metaphor in Renaissance English Literature*. The Hague: Mouton, 1971.

Hall, Thomas S. *Ideas of Life and Matter: Studies in the History of General Physiology*. 2 vols. Chicago: University of Chicago Press, 1969.

Halpern, Richard. "Puritanism and Maenadism in *A Mask*." In *Rewriting the Renaissance: The Discourses of Sexual Difference in Early Modern Europe*, ed. Margaret W. Ferguson, Maureen Quilligan, and Nancy J. Vickers, pp. 88–105. Chicago: University of Chicago Press, 1986.

Hammond, Gerald. *Fleeting Things: English Poets and Poems, 1616–1660*. Cambridge: Harvard University Press, 1990.

Harrison, Charles. "Ancient Atomists and English Literature of the Seventeenth Century." *Harvard Studies in Classical Philology* 45 (1934): 1–79.

Hartman, Geoffrey. *Beyond Formalism: Literary Essays, 1958–1970*. New Haven: Yale University Press, 1970.

Hartman, Mark. "Hobbes's Concept of Political Revolution." *Journal of the History of Ideas* 47 (1986): 487–95.

Harvey, Elizabeth D., and Katharine Eisaman Maus, eds. *Soliciting Interpretation: Literary Theory and Seventeenth-Century English Poetry*. Chicago: University of Chicago Press, 1990.

Hayes, T. Wilson. *Winstanley the Digger: A Literary Analysis of Radical Ideas in the English Revolution*. Cambridge: Harvard University Press, 1979.

Hayes, Thomas. "Dialectical Development in Andrew Marvell's Life, Poetry, and Historical Milieu." Ph.D. dissertation, New York University, 1970.

Heimann, P. M. "Voluntarism and Immanence: Conceptions of Nature in Eighteenth-Century Thought." *Journal of the History of Ideas* 39 (1978): 271–83.

Henry, John. "The Matter of Souls: Medical Theory and Theology in Seventeenth-Century England." In *The Medical Revolution of the Seventeenth Century*, ed. Roger French and Andrew Wear, pp. 87–113. Cambridge: Cambridge University Press, 1989.

———. "Medicine and Pneumatology: Henry More, Richard Baxter, and Francis Glisson's *Treatise on the Energetic Nature of Substance*." *Medical History* 31 (1987): 15–40.

———. "Occult Qualities and the Experimental Philosophy: Active Principles in Pre-Newtonian Matter Theory." *History of Science* 24 (1986): 335–81.

Herz, Judith Scherer. "Milton and Marvell: The Poet as Fit Reader." *Modern Language Quarterly* 39 (1978): 239–63.

Herzog, Don. *Happy Slaves: A Critique of Consent Theory*. Chicago: University of Chicago Press, 1989.

Hill, Christopher. *The Experience of Defeat: Milton and Some Contemporaries*. New York: Penguin, 1985.

———. *God's Englishman: Oliver Cromwell and the English Revolution*. New York: Harper, 1970.

———. *Milton and the English Revolution*. Harmondsworth: Penguin, 1977.

———. *Puritanism and Revolution*. London: Secker and Warburg, 1958.

———. *The Religion of Gerrard Winstanley*. *Past and Present*, supplement no. 5, 1978.

———. "Science and Magic in Seventeenth-Century England." In *Culture, Ideology, and Politics*, ed. Raphael Samuel and Gareth Stedman Jones, pp. 176–93. London: Routledge, 1982.

———. *Society and Puritanism in Pre-Revolutionary England*. London: Continuum, 1964.

———. "William Harvey and the Idea of Monarchy." *Past and Present* 27 (1964). Reprinted in *The Intellectual Revolution of the Seventeenth Century*, ed. Charles Webster, pp. 160–81. London: Routledge, 1974.

———. "The Word 'Revolution' in Seventeenth-Century England." In *For Veronica Wedgwood These: Studies in Seventeenth-Century History*, ed. Richard Ollard and Pamela Tudor-Craig, pp. 134–51. London: Collins, 1986.

———. *The World Turned Upside Down: Radical Ideas during the English Revolution*. Harmondsworth: Penguin, 1975.

Hodge, R. I. V. *Foreshortened Time: Andrew Marvell and Seventeenth-Century Revolutions*. Cambridge: Brewer, 1978.

———. "Satan and the Revolution of the Saints." In *Literature, Language, and Society in England, 1580–1680*, ed. David Aers, Bob Hodge, and Gunther Kress. Dublin: Gill and Macmillan, 1981.

Hollander, John. *The Figure of Echo: A Mode of Allusion in Milton and After*. Berkeley: University of California Press, 1981.

———. *The Untuning of the Sky: Ideas of Music in English Poetry, 1500–1700*. Princeton: Princeton University Press, 1961.

Hollington, Michael, ed. *Paradise Lost, Books XI–XII*. Cambridge: Cambridge University Press, 1976.

Holstun, James. *A Rational Millennium: Puritan Utopias of Seventeenth-Century England and America*. New York: Oxford University Press, 1987.

Howard, Leon. "'The Invention' of Milton's 'Great Argument': A Study of the Logic of 'God's Ways to Men.'" *Huntington Library Quarterly* 9 (1945–46): 149–73.

Hudson, Winthrop S. "Economic and Social Thought of Gerrard Winstanley: Was He a Seventeenth-Century Marxist?" *Journal of Modern History* 18 (1946): 1–21.

Hughes, Merritt Y. "Milton and the Sense of Glory." *Philological Quarterly* 28 (1949): 107–24.

———, gen. ed. *A Variorum Commentary on the Poems of John Milton*. New York: Columbia University Press, 1970–72.

Hunter, G. K. *Paradise Lost*. Unwin Critical Library Series. London: Allen and Unwin, 1980.

Hunter, Michael. *Establishing the New Science: The Experience of the Early Royal Society*. Wolfeboro, N. H.: Boydell and Brewer, 1989.

——. *Science and Society in Restoration England*. Cambridge: Cambridge University Press, 1981.

——, ed. *Robert Boyle Reconsidered*. Cambridge: Cambridge University Press, 1994.

Hunter, William B., Jr. "Milton and Thrice-Great Hermes." *Journal of English and Germanic Philology* 45 (1946): 327–36.

——. "Milton's Materialistic Life Principle." *Journal of English and Germanic Philology* 45 (1946): 68–76.

——. "Milton's Power of Matter." *Journal of the History of Ideas* 13 (1952): 551–62.

Hunter, William B., Jr., C. A. Patrides, and J. H. Adamson, eds. *Bright Essence: Studies in Milton's Theology*. Salt Lake City: University of Utah Press, 1971.

Hutton, Sarah, ed. *Henry More (1614–1687) Tercentenary Studies*. Dordrecht: Kluwer, 1990.

Jacob, James R. "Boyle's Atomism and the Restoration Assault on Pagan Naturalism." *Social Studies of Science* 8 (1978): 211–34.

——. *Henry Stubbe, Radical Protestantism, and the Early Enlightenment*. Cambridge: Cambridge University Press, 1983.

——. "The Ideological Origins of Robert Boyle's Natural Philosophy." *Journal of European Studies* 2 (1972): 1–21.

——. *Robert Boyle and the English Revolution: A Study in Social and Intellectual Change*. New York: Franklin, 1977.

Jacob, James R., and Margaret C. Jacob. "The Anglican Origins of Modern Science: The Metaphysical Foundations of the Whig Constitution." *Isis* 71 (1980): 251–67.

Jacob, Margaret C. *The Cultural Meaning of the Scientific Revolution*. Philadelphia: Temple University Press, 1988.

——. *The Radical Enlightenment: Pantheists, Freemasons, and Republicans*. London: Allen and Unwin, 1981.

Jameson, Fredric. "Religion and Ideology: A Political Reading of *Paradise Lost*." In *Literature, Politics, and Theory*, ed. Francis Barker, Peter Hulme, Margaret Iversen, and Diana Loxley, pp. 35–56. London: Methuen, 1986.

Johnson, James Turner. *A Society Ordained by God: English Puritan Marriage Doctrine in the First Half of the Seventeenth Century*. Nashville, Tenn.: Abingdon, 1970.

Jones, Kathleen. *A Glorious Fame: The Life of Margaret Cavendish, Duchess of Newcastle, 1623–1673*. London: Bloomsbury, 1988.

Kahn, Victoria. "Allegory and the Sublime in *Paradise Lost*." In *John Milton*, ed. Annabel Patterson, pp. 185–201. London: Longman, 1992.

——. *Machiavellian Rhetoric: From the Counter-Reformation to Milton*. Princeton: Princeton University Press, 1994.

——. *Rhetoric, Prudence, and Scepticism in the Renaissance*. Ithaca: Cornell University Press, 1985.

Kargon, Robert Hugh. *Atomism in England from Hariot to Newton*. Oxford: Clarendon, 1966.

Kegl, Rosemary. " 'Joyning my Labour to my Pain': The Politics of Labor in Marvell's Mower Poems." In *Soliciting Interpretation: Literary Theory and Seventeenth-Century English Poetry*, ed. Elizabeth D. Harvey and Katharine Eisaman Maus, pp. 89–118. Chicago: University of Chicago Press, 1990.

Keller, Evelyn Fox. *Reflections on Gender and Science*. New Haven: Yale University Press, 1985.

Kendrick, Christopher. *Milton: A Study in Ideology and Form*. New York: Methuen, 1986.

———. "Milton and Sexuality: A Symptomatic Reading of *Comus*." In *Re-membering Milton: Essays on the Texts and Traditions*, ed. Mary Nyquist and Margaret W. Ferguson, pp. 43–73. New York: Methuen, 1987.

Kerrigan, William. "The Heretical Milton: From Assumption to Mortalism." *English Literary Renaissance* 5 (1975): 125–66.

———. "Marvell and Nymphets." *Greyfriar* 27 (1986): 3–21.

———. "The Politically Correct *Comus*: A Reply to John Leonard." *Milton Quarterly* 27 (1993): 149–53.

———. *The Prophetic Milton*. Charlottesville: University Press of Virginia, 1974.

———. *The Sacred Complex: On the Psychogenesis of "Paradise Lost."* Cambridge: Harvard University Press, 1983.

Keynes, Geoffrey. *A Bibliography of the Writings of Dr. William Harvey, 1578–1657*. 2d ed. Cambridge: Cambridge University Press, 1953.

———. *The Life of William Harvey*. Oxford: Clarendon, 1966.

Klause, John. *The Unfortunate Fall: Theodicy and the Moral Imagination of Andrew Marvell*. Hamden, Conn.: Archon, 1983.

Korshin, Paul J., ed. *Studies in Change and Revolution: Aspects of English Intellectual History, 1640–1800*. Menston, England: Scolar Press, 1972.

Kroll, Richard W. F. *The Material Word: Literate Culture in the Restoration and Early Eighteenth Century*. Baltimore: Johns Hopkins University Press, 1991.

Kroll, Richard W. F., Richard Ashcraft, and Perez Zagorin, eds. *Philosophy, Science, and Religion in England, 1640–1700*. Cambridge: Cambridge University Press, 1992.

Lamprecht, Sterling. "The Role of Descartes in Seventeenth-Century England." *Studies in the History of Ideas* 3 (1935): 184–240.

Lawrence, Karen R. *Penelope Voyages: Women and Travel in the British Literary Tradition*. Ithaca: Cornell University Press, 1994.

le Goff, Jacques. *Time, Work, and Culture in the Middle Ages*. Trans. Arthur Goldhammer. Chicago: University of Chicago Press, 1980.

Leishman, J. B. *The Art of Marvell's Poetry*. London: Hutchinson and Co., 1966.

Leonard, John. "Saying 'No' to Freud: Milton's *A Mask* and Sexual Assault." *Milton Quarterly* 25 (1991): 129–40.

Lewalski, Barbara K. *Donne's "Anniversaries" and the Poetry of Praise: The Creation of a Symbolic Mode*. Princeton: Princeton University Press, 1973.

———. "Milton: Political Beliefs and Polemical Methods, 1659–60." *PMLA* 74 (1959): 191–202.

———. *"Paradise Lost" and the Rhetoric of Literary Forms*. Princeton: Princeton University Press, 1985.

———. "Structure and the Symbolism of Vision in Michael's Prophecy, *Paradise Lost*, XI–XII." *Philological Quarterly* 42 (1963): 23–35.

———. "Typology and Poetry: A Consideration of Herbert, Vaughan, and Marvell." In

Illustrious Evidence: Approaches to English Literature of the Early Seventeenth Century, ed. Earl Miner, pp. 41–70. Berkeley: University of California Press, 1975.

Lewis, C. S. *A Preface to "Paradise Lost."* London: Oxford University Press, 1942.

Lieb, Michael. *The Dialectics of Creation: Patterns of Birth and Regeneration in "Paradise Lost."* Amherst: University of Massachusetts Press, 1970.

——. "Further Thoughts on Satan's Journey through Chaos." *Milton Quarterly* 12 (1978): 126–33.

——. *Milton and the Culture of Violence*. Ithaca: Cornell University Press, 1994.

——. *Poetics of the Holy: A Reading of "Paradise Lost."* Chapel Hill: University of North Carolina Press, 1981.

——. *The Sinews of Ulysses: Form and Convention in Milton's Works*. Pittsburgh: Duquesne University Press, 1988.

Loewenstein, David A. "*Areopagitica* and the Dynamics of History." *Studies in English Literature* 28 (1988): 77–93.

——. *Milton and the Drama of History: Historical Vision, Iconoclasm, and the Literary Imagination*. Cambridge: Cambridge University Press, 1990.

Lord, George deF. *Classical Presences in Seventeenth-Century Poetry*. New Haven: Yale University Press, 1989.

——, ed. *Andrew Marvell: A Collection of Critical Essays*. Twentieth Century Views Series. Englewood Cliffs, N.J.: Prentice-Hall, 1968.

Low, Anthony. *The Georgic Revolution*. Princeton: Princeton University Press, 1985.

Lynch, William T. "Politics in Hobbes' Mechanics: The Social as Enabling." *Studies in History and Philosophy of Science* 22 (1991): 295–320.

MacCaffrey, Isabel Gamble. *"Paradise Lost" as "Myth."* Cambridge: Harvard University Press, 1959.

McColley, Diane Kelsey. *Milton's Eve*. Urbana: University of Illinois Press, 1983.

McGuire, Mary Ann. "Margaret Cavendish, Duchess of Newcastle, on the Nature and Status of Women." *International Journal of Women's Studies* 1 (1978): 193–206.

MacLean, Gerald M. *Time's Witness: Historical Representation in English Poetry, 1603–1660*. Madison: University of Wisconsin Press, 1990.

Maclean, Ian. *The Renaissance Notion of Woman: A Study in the Fortunes of Scholasticism and Medical Science in European Intellectual Life*. Cambridge: Cambridge University Press, 1980.

Macpherson, C. B. *The Political Theory of Possessive Individualism: Hobbes to Locke*. Oxford: Clarendon, 1962.

Madsen, William G. *From Shadowy Types to Truth: Studies in Milton's Symbolism*. New Haven: Yale University Press, 1968.

Marcus, Leah S. *Childhood and Cultural Despair: A Theme and Variations in Seventeenth-Century English Literature*. Pittsburgh: University of Pittsburgh Press, 1978.

——. *The Politics of Mirth: Jonson, Herrick, Milton, Marvell, and the Defense of Old Holiday Pastimes*. Chicago: University of Chicago Press, 1986.

Marjara, Harinder Singh. *Contemplation of Created Things: Science in "Paradise Lost."* Toronto: University of Toronto Press, 1992.

Martz, Louis. *The Paradise Within: Studies in Vaughan, Traherne, and Milton*. New Haven: Yale University Press, 1980.

———. *Poet of Exile: A Study of Milton's Poetry*. New Haven: Yale University Press, 1980.

Masson, David. *The Life of John Milton*. 5 vols. London: Macmillan, 1877.

Mayr, Otto. *Authority, Liberty, and Automatic Machinery in Early Modern Europe*. Baltimore: Johns Hopkins University Press, 1986.

Mendelsohn, J. Andrew. "Alchemy and Politics in England, 1649–1665." *Past and Present* 135 (1992): 30–78.

Mendelson, Sara Heller. *The Mental World of Stuart Women: Three Studies*. Amherst: University of Massachusetts Press, 1987.

Merchant, Carolyn. *The Death of Nature: Women, Ecology, and the Scientific Revolution*. San Francisco: Harper and Row, 1980.

Meyer, Arthur William. *An Analysis of the "De Generatione Animalium" of William Harvey*. Stanford: Stanford University Press, 1936.

Milner, Andrew. *John Milton and the English Revolution: A Study in the Sociology of Literature*. London: Macmillan, 1981.

Mintz, Samuel I. "The Duchess of Newcastle's Visit to the Royal Society." *Journal of English and Germanic Philology* 5 (1952): 168–76.

———. *The Hunting of Leviathan: Seventeenth-Century Reactions to the Materialism and Moral Philosophy of Thomas Hobbes*. Cambridge: Cambridge University Press, 1962.

Mulder, David. *The Alchemy of Revolution: Gerrard Winstanley's Occultism and Seventeenth-Century English Communism*. New York: Lang, 1990.

Nicolson, Marjorie Hope. "Early Stage of Cartesianism in England." *Studies in Philosophy* 26 (1929): 451–74.

———. "Milton and Hobbes." *Studies in Philology* 23 (1926): 405–33.

———. "The Spirit World of Milton and More." *Studies in Philology* 22 (1925): 433–52.

Norbrook, David. "*Areopagitica*, Censorship, and the Early Modern Public Sphere." In *The Administration of Aesthetics: Censorship, Political Criticism, and the Public Sphere*, ed. Richard Burt, pp. 3–33. Minneapolis: University of Minnesota Press, 1994.

———. *Poetry and Politics in the English Renaissance*. London: Routledge and Kegan Paul, 1984.

Nyquist, Mary. "Gynesis, Genesis, Exegesis, and the Formation of Milton's Eve." In *Cannibals, Witches, and Divorce*, ed. Marjorie Garber, pp. 147–208. Baltimore: Johns Hopkins University Press, 1987.

Nyquist, Mary, and Margaret Ferguson, eds. *Re-membering Milton: Essays on the Texts and Traditions*. New York: Methuen, 1988.

Oakley, Francis. *Omnipotence, Covenant, and Order: An Excursion in the History of Ideas from Abelard to Leibniz*. Ithaca: Cornell University Press, 1984.

Osler, Margaret J. *Divine Will and the Mechanical Philosophy: Gassendi and Descartes on Contingency and Necessity in the Created World*. Cambridge: Cambridge University Press, 1994.

———. "The Intellectual Sources of Robert Boyle's Philosophy of Nature: Gassendi's

Voluntarism and Boyle's Physico-Theological Project." In *Philosophy, Science, and Religion in England, 1640–1700*, ed. Richard Kroll, Richard Ashcraft, and Perez Zagorin, pp. 178–98. Cambridge: Cambridge University Press, 1992.

Osler, Margaret J., and Paul Lawrence Farber, eds. *Religion, Science, and Worldview: Essays in Honor of Richard S. Westfall.* Cambridge: Cambridge University Press, 1985.

Oster, Malcolm. "Virtue, Providence, and Political Neutralism: Boyle and Interregnum Politics." In *Robert Boyle Reconsidered*, ed. Michael Hunter, pp. 19–36. Cambridge: Cambridge University Press, 1994.

Pagel, Walter. *From Paracelsus to Van Helmont: Studies in Renaissance Medicine and Science.* Ed. Marianne Winder. London: Variorum Reprints, 1986.

——. *Joan Baptista Van Helmont: Reformer of Science and Medicine.* Cambridge: Cambridge University Press, 1982.

——. *New Light on William Harvey.* Basel: Karger, 1976.

——. *Paracelsus: An Introduction to Philosophical Medicine in the Era of the Renaissance.* 2d ed. Basel: Karger, 1982.

——. *Religion and Neoplatonism in Renaissance Medicine.* Ed. Marianne Winder. London: Variorum Reprints, 1985.

——. *The Religious and Philosophical Aspects of van Helmont's Science and Medicine.* Baltimore: Johns Hopkins University Press, 1944.

——. *William Harvey's Biological Ideas: Selected Aspects and Historical Background.* New York: Hafner, 1967.

Pagels, Elaine. *Adam, Eve, and the Serpent.* New York: Random House, 1988.

Parker, W. R. *Milton: A Biography.* 2 vols. Oxford: Clarendon, 1968.

Parry, Graham. *Seventeenth-Century Poetry: The Social Context.* London: Hutchinson, 1985.

Patrides, C. A. *The Grand Design of God: The Literary Form of the Christian View of History.* London: Routledge and Kegan Paul; Toronto: University of Toronto Press, 1972.

——. *Milton and the Christian Tradition.* Oxford: Clarendon, 1966.

——. *The Phoenix and the Ladder: The Rise and Decline of the Christian View of History.* Berkeley: University of California Press, 1964.

——. " 'Something Like Prophetic Strain': Apocalyptic Configurations in Milton." *English Language Notes* 19, no. 3 (1982): 3–25.

Patrides, C. A., and R. B. Waddington, eds. *The Age of Milton: Backgrounds to Seventeenth-Century Literature.* Manchester: Manchester University Press, 1980.

Patrides, C. A., and Joseph Wittreich, eds. *The Apocalypse in English Renaissance Thought and Literature.* Ithaca: Cornell University Press, 1984.

Patterson, Annabel. " 'Bermudas' and 'The Coronet': Marvell's Protestant Ethics." *ELH* 44 (1977): 478–99.

——. *Censorship and Interpretation: The Condition of Writing and Reading in Early Modern England.* Madison: University of Wisconsin Press, 1984.

——. *Fables of Power: Aesopian Writing and Political History.* Durham: Duke University Press, 1991.

——. *Marvell and the Civic Crown.* Princeton: Princeton University Press, 1978.

——. *Pastoral and Ideology: Virgil to Valéry*. Berkeley: University of California Press, 1987.

——. "Pastoral versus Georgic: The Politics of Virgilian Quotation." In *Renaissance Genres: Essays on Theory, History, and Interpretation*, ed. Barbara Kiefer Lewalski, pp. 241–67. Cambridge: Harvard University Press, 1986.

——. *Reading between the Lines*. Madison: University of Wisconsin Press, 1993.

Perry, Henry Ten Eyck. *The First Duchess of Newcastle and Her Husband as Figures in Literary History*. Boston: Ginn and Co., 1918.

Petegorsky, David W. *Left-Wing Democracy in the English Civil War: A Study of the Social Philosophy of Gerrard Winstanley*. London: Gollancz, 1940.

Pocock, J. G. A. *The Ancient Constitution and the Feudal Law: A Study of English Historical Thought in the Seventeenth Century*. 2d ed. Cambridge: Cambridge University Press, 1987.

——. *The Machiavellian Moment: Florentine Political Thought and the Atlantic Republican Tradition*. Princeton: Princeton University Press, 1975.

Poynter, F. N. L. "Nicholas Culpepper and the Paracelsians." In *Science, Medicine, and Society in the Renaissance: Essays to Honor Walter Pagel*, ed. Allen G. Debus, 2 vols., 1:201–20. New York: Watson, 1972.

Quint, David. *Epic and Empire: Politics and Generic Form from Virgil to Milton*. Princeton: Princeton University Press, 1993.

——. *From Origin to Originality: Versions of the Source in Renaissance Literature*. Princeton: Princeton University Press, 1980.

Radzinowicz, Mary Ann. "Man as a Probationer of Immortality: *Paradise Lost* XI–XII." In *Approaches to "Paradise Lost": The York Tercentenary Lectures*, ed. C. A. Patrides, pp. 31–51. Toronto: University of Toronto Press, 1968.

——. *Milton's Epics and the Book of Psalms*. Princeton: Princeton University Press, 1989.

——. "The Politics of *Paradise Lost*." In *Politics of Discourse: The Literature and History of Seventeenth-Century England*, ed. Kevin Sharpe and Steven N. Zwicker, pp. 203–29. Berkeley: University of California Press, 1987.

——. *Toward "Samson Agonistes": The Growth of Milton's Mind*. Princeton: Princeton University Press, 1978.

Rajan, Balachandra. "*Paradise Lost*: The Hill of History." *Huntington Library Quarterly* 31 (1967): 43–63.

——. "*Paradise Lost*" and the Seventeenth-Century Reader". London: Chatto and Windus, 1947.

Rattansi, P. M. "Paracelsus and the Puritan Revolution." *Ambix* 11 (1963): 24–32.

——. "The Social Interpretation of Science in the Seventeenth Century." In *Science and Society, 1600–1900*, ed. Peter Mathias, pp. 1–32. Cambridge: Cambridge University Press, 1972.

Reesing, John. "The Materiality of God in Milton's *De Doctrina Christiana*." *Harvard Theological Review* 50 (1957): 159–74.

Revard, Stella P. *The War in Heaven: "Paradise Lost" and the Tradition of Satan's Rebellion*. Ithaca: Cornell University Press, 1980.

Rogers, John. "The Enclosure of Virginity: The Poetics of Sexual Abstinence in the

English Revolution." In *Enclosure Acts: Sexuality, Property, and Culture in Early Modern England*, ed. Richard Burt and John Michael Archer, pp. 229–50. Ithaca: Cornell University Press, 1994.

Romanell, Patrick. *John Locke and Medicine: A New Key to Locke*. Buffalo, New York: Prometheus, 1984.

Rosenblatt, Jason P. "Adam's Pisgah Vision: *Paradise Lost*, Books XI and XII." *ELH* 39 (1972): 66–86.

———. *Torah and Law in "Paradise Lost."* Princeton: Princeton University Press, 1994.

Rossi, Paolo. *The Dark Abyss of Time: The History of the Earth and The History of Nations from Hooke to Vico*. Trans. Lydia G. Cochrane. Chicago: University of Chicago Press, 1984.

Røstvig, Maren-Sofie. *The Happy Man: Studies in the Metamorphoses of a Classical Ideal*. 2 vols. Oxford: Blackwell, 1954–58.

———. " 'Upon Appleton House' and the Universal History of Man." *English Studies* (Netherlands) 42 (1961): 337–51.

Rumrich, John Peter. *Matter of Glory: A New Preface to "Paradise Lost."* Pittsburgh: University of Pittsburgh Press, 1987.

———. "Uninventing Milton." *Modern Philology* 87 (1990): 249–65.

Sarasohn, Lisa T. "Motion and Morality: Pierre Gassendi, Thomas Hobbes, and the Mechanical World-View." *Journal of the History of Ideas* 46 (1985): 363–79.

———. "A Science Turned Upside Down: Feminism and the Natural Philosophy of Margaret Cavendish." *Huntington Library Quarterly* 47 (1984): 289–307.

Saurat, Denis. *Milton: Man and Thinker*. New York: Dial Press, 1925.

Savage, J. B. "Freedom and Necessity in *Paradise Lost*." *ELH* 44 (1977): 286–311.

Schaffer, Simon. "Godly Men and Mechanical Philosophers: Souls and Spirits in Restoration Natural Philosophy." *Science in Context* 1 (1987): 55–85.

Schiebinger, Londa. *The Mind Has No Sex? Women in the Origins of Modern Science*. Cambridge: Harvard University Press, 1989.

Schochet, Gordon. *Patriarchalism in Political Thought: The Authoritarian Family and Political Speculation and Attitudes Especially in Seventeenth-Century England*. Oxford: Blackwell, 1975.

Schwartz, Regina M. *Remembering and Repeating: On Milton's Theology and Poetics*. Chicago: University of Chicago Press, 1988.

Scoular, Kitty W. *Natural Magic*. Oxford: Clarendon, 1965.

Sennett, Richard. *Flesh and Stone: The Body and the City in Western Civilization*. New York: Norton, 1994.

Sewell, Arthur. *A Study in Milton's Christian Doctrine*. Oxford: Oxford University Press, 1939.

Shapin, Steven. "Of Gods and Kings: Natural Philosophy and Politics in the Leibniz-Clarke Disputes." *Isis* 72 (1981): 187–215.

———. "Social Uses of Science." In *The Ferment of Knowledge: Studies in the Historiography of Eighteenth-Century Science*, ed. G. S. Rousseau and Roy Porter, pp. 95-139. Cambridge: Cambridge University Press, 1980.

Shapin, Steven, and Simon Schaffer. *Leviathan and the Air-Pump: Hobbes, Boyle, and the Experimental Life*. Princeton: Princeton University Press, 1985.

Shapiro, Barbara J. *Probability and Certainty in Seventeenth-Century England: A Study of the Relationships between Natural Science, Religion, History, Law, and Literature*. Princeton: Princeton University Press, 1983.

Sharpe, Kevin, and Steven N. Zwicker, eds. *Politics of Discourse: The Literature and History of Seventeenth-Century England*. Berkeley: University of California Press, 1987.

Shawcross, John T. *With Mortal Voice: The Creation of "Paradise Lost."* Lexington: University Press of Kentucky, 1982.

Shulman, George M. *Radicalism and Reverence: The Political Thought of Gerrard Winstanley*. Berkeley: University of California Press, 1989.

Sirluck, Ernest. "Milton's Political Thought: The First Cycle." *Modern Philology* 61 (1964): 209–24.

Smith, Hilda. *Reason's Disciples: Seventeenth-Century English Feminists*. Urbana: University of Illinois Press, 1982.

Smith, Nigel. *Literature and Revolution in England, 1640–1660*. New Haven: Yale University Press, 1994.

——. *Perfection Proclaimed: Language and Literature in English Radical Religion, 1640–1660*. Oxford: Clarendon, 1989.

Stallybrass, Peter. "Shakespeare, the Individual, and the Text." In *Cultural Studies*, ed. Lawrence Grossberg, Cary Nelson, and Paula A. Treichler, pp. 593-610. New York: Routledge, 1992.

Steadman, John. *Milton's Epic Characters: Image and Idol*. Chapel Hill: University of North Carolina Press, 1968.

Stocker, Margarita. *Apocalyptic Marvell: The Second Coming in Seventeenth-Century Poetry*. Sussex: Harvester, 1986.

Summers, Claude J. "The Frightened Architects of Marvell's 'Horatian Ode.'" *Seventeenth-Century News* 28 (1970): 5.

Summers, Claude J., and Ted-Larry Pebworth, eds. *On the Celebrated and Neglected Poems of Andrew Marvell*. Columbia: University of Missouri Press, 1992.

Summers, Joseph H. "The Final Vision." In *Milton: A Collection of Critical Essays*, ed. Louis Martz, pp. 183–206. Englewood Cliffs, N.J.: Prentice-Hall, 1966.

——. "Marvell's 'Nature.'" *ELH* 20 (1953): 121–35.

——. "Some Apocalyptic Strains in Marvell's Poetry." In *Tercentenary Essays in Honour of Andrew Marvell*, ed. Kenneth Friedenreich, pp. 180–203. Hamden, Conn.: Archon, 1977.

Svendson, Kester. *Milton and Science*. Cambridge: Harvard University Press, 1956.

Swaim, Kathleen. *Before and After the Fall*. Amherst: University of Massachusetts Press, 1986.

Swan, Jim. "At Play in the Garden of Ambivalence: Andrew Marvell and the Green World." *Criticism* 17 (1975): 295–307.

——. "Difference and Silence: John Milton and the Question of Gender." In *The (M)other Tongue: Essays in Feminist Psychoanalytic Interpretation*, ed. Shirley Nelson Garner, Claire Kahane, and Madelon Sprengnether, pp. 142–68. Ithaca: Cornell University Press, 1985.

——. "History, Pastoral, and Desire: Andrew Marvell's Mower Poems." *International Review of Psycho-Analysis* 3 (1976): 193–202.

Tayler, Edward. *Milton's Poetry: Its Development in Time*. Pittsburgh: Duquesne University Press, 1979.

Thomas, Keith. *Man and the Natural World: A History of the Modern Sensibility*. New York: Pantheon, 1983.

———. *Religion and the Decline of Magic*. New York: Scribner's, 1971.

———. "Women and the Civil War Sects." *Past and Present* 13 (1958): 42–62.

Toliver, Harold. *Marvell's Ironic Vision*. New Haven: Yale University Press, 1965.

Trevor-Roper, Hugh. "The Paracelsian Movement." In his *Renaissance Essays*, pp. 149–99. London: Secker and Warburg, 1985.

Trompf, G. W. *The Idea of Historical Recurrence in Western Thought: From Antiquity to the Reformation*. Berkeley: University of California Press, 1979.

Tuck, Richard. *Hobbes*. Oxford: Oxford University Press, 1989.

———. *Natural Rights Theories: Their Origin and Development*. Cambridge: Cambridge University Press, 1979.

Turner, James Grantham. *One Flesh: Paradisal Marriage and Sexual Relations in the Age of Milton*. Oxford: Clarendon, 1987.

———. *The Politics of Landscape: Rural Scenery and Society in English Poetry, 1630–1660*. Cambridge: Harvard University Press, 1979.

Waldock, A. J. A. *"Paradise Lost" and Its Critics*. Gloucester, Mass.: Peter Smith, 1959.

Wallace, John M. *Destiny His Choice: The Loyalism of Andrew Marvell*. Cambridge: Cambridge University Press, 1968.

Walzer, Michael. *The Revolution of the Saints: A Study in the Origins of Radical Politics*. New York: Atheneum, 1976.

Watkins, W. B. C. *An Anatomy of Milton's Verse*. Baton Rouge: Louisiana State University Press, 1955.

Webster, Charles. "English Medical Reformers of the Puritan Revolution: A Background to the 'Society of Chymical Physitians.'" *Ambix* 14 (1967): 16–41.

———. *From Paracelsus to Newton: Magic and the Making of Modern Science*. Cambridge: Cambridge University Press, 1980.

———. *The Great Instauration: Science, Medicine, and Reform, 1626–1660*. London: Duckworth, 1975.

———. "Henry Power's Experimental Philosophy." *Ambix* 14 (1967): 150–78.

———, ed. *The Intellectual Revolution of the Seventeenth Century*. London: Routledge, 1974.

Westfall, Richard. *Science and Religion in Seventeenth-Century England*. New Haven: Yale University Press, 1958.

Wheeler, L. Richmond. *Vitalism: Its History and Validity*. London: Witherby, 1939.

Whitteridge, Gweneth. "William Harvey: A Royalist and No Parliamentarian." *Past and Present* 30 (1964): 104–109. Reprinted in *The Intellectual Revolution of the Seventeenth Century*, ed. Charles Webster, pp. 182–88. London: Routledge, 1974.

———. *William Harvey and the Circulation of the Blood*. London: Macdonald, 1971.

Wilcher, Robert. *Andrew Marvell*. Cambridge: Cambridge University Press, 1985.

Wilding, Michael. *Dragons Teeth: Literature in the English Revolution*. Oxford: Clarendon, 1987.

——, ed. *Marvell: Modern Judgements*. London: Macmillan, 1969.

Willey, Basil. *The Seventeenth-Century Background: Studies in the Thought of the Age in Relation to Poetry and Religion*. New York: Doubleday Anchor Books, 1934.

Williams, Raymond. *The Country and the City*. New York: Oxford University Press, 1973.

——. *Keywords: A Vocabulary of Culture and Society*. Revised edition. New York: Oxford University Press, 1983.

Williamson, George. "Milton and the Mortalist Heresy." *Studies in Philology* 32 (1935): 553–79.

Williamson, Marilyn. *Raising their Voices: British Women Writers, 1650–1750*. Detroit: Wayne State University Press, 1990.

Wittreich, Joseph Anthony, Jr., ed. *Calm of Mind: Tercentenary Essays on "Paradise Regained" and "Samson Agonistes" in Honor of John S. Diekhoff*. Cleveland: Case Western Reserve University Press, 1971.

——. *Feminist Milton*. Ithaca: Cornell University Press, 1987.

——. *Visionary Poetics: Milton's Tradition and His Legacy*. San Marino, Calif.: Huntington Library Press, 1979.

Wolfe, Don M. *Milton in the Puritan Revolution*. 1941. New York: Humanities Press, 1963.

Woodhouse, A. S. P. "Notes on Milton's Views on the Creation: The Initial Phases." *Philological Quarterly* 27 (1949): 211–36.

Woolf, Virginia. *The Common Reader*. 1925. New York: Harcourt, Brace, 1948.

Worden, Blair. "Providence and Politics in Cromwellian England." *Past and Present* 109 (1985): 54–97.

Yolton, John. *Thinking Matter: Materialism in Eighteenth-Century Britain*. Oxford: Oxford University Press, 1984.

Zwicker, Steven N. *Lines of Authority: Politics and English Literary Culture, 1649–1689*. Ithaca: Cornell University Press, 1993.

INDEX